Mössbauer Effect Methodology

Volume 5

MÖSSBAUER EFFECT METHODOLOGY

Proceedings of Annual Symposia Held in New York City, Sponsored by the New England Nuclear Corporation, Boston

Edited by Irwin J. Gruverman

Volume 1
Proceedings of the First Symposium–January 1965

Volume 2
Proceedings of the Second Symposium–January 1966

Volume 3
Proceedings of the Third Symposium–January 1967

Volume 4
Proceedings of the Fourth Symposium–January 1968

Volume 5
Proceedings of the Fifth Symposium–February 1969

A Publication of the New England Nuclear Corporation

Mössbauer Effect Methodology

Volume 5

Proceedings of the Fifth Symposium on
Mössbauer Effect Methodology
New York City, February 2, 1969

Edited by
Irwin J. Gruverman
Consultant, Special Sources Department
New England Nuclear Corporation
Boston, Massachusetts

PLENUM PRESS • NEW YORK – LONDON • 1970

Library of Congress Catalog Card Number 65-21188

SBN 306-38805-7

A Division of Plenum Publishing Corporation
227 West 17th Street, New York, N.Y. 10011

United Kingdom edition published by Plenum Press, London
A Division of Plenum Publishing Company, Ltd.
Donington House, 30 Norfolk Street, London W.C.2, England

Printed in the United States of America

Preface

This is the fifth volume of a series which provides a continuing forum for publication of developments in Mössbauer effect methodology and of spectroscopy and its applications.

Mössbauer Effect Methodology, Volume 5, records the proceedings of the Fifth Symposium on Mössbauer Effect Methodology. The symposium was sponsored by the New England Nuclear Corporation and was centered on the themes of spectroscopy, new applications, and methodology. The symposium was held in the Mercury Ballroom of the New York Hilton hotel on February 2, 1969. Dr. P. A. Flinn of Carnegie-Mellon Institute was chairman of the afternoon and evening sessions.

About three hundred participants attended, and this degree of interest leads us to anticipate a sixth symposium early in 1970.

Elron Electronic Industries and Reuter-Stokes Electronic Components Company demonstrated lines of equipment for Mössbauer investigators. The evident high quality of the commercial instrumentation available is a tribute to the growth of Mössbauer technology and to the manufacturers.

It was quite difficult to select the 14 papers for the symposium from the many submitted. We wish to express our gratitude to all of the authors who submitted papers, and particularly to those we were forced to disappoint, for their understanding of our limitations. This symposium concentrated on spectroscopy and applications, with some emphasis on methodology. Environmental control with a He^3/He^4 refrigerator and techniques for measurements in radioactive materials were of particular interest. Many of the papers report work in areas remote from pure physics, and include metallurgical, minerological, and biological applications.

The editor is indebted to his colleagues for their assistance in recommending and selecting papers, to the authors for their excellent work, and very much to Paul Flinn for his skilled handling of a long and interesting program.

I. G.

Needham, Massachusetts
May, 1969

Contents

Spectroscopy

Applications

Methodology

Spectroscopy

Mössbauer Effect Studies of Lattice-Dynamic Anisotropy and Line Asymmetry in Semiconductor and Organometallic Tin Compounds

Hans A. Stöckler

Department of Chemical Physics
National Lead Company Research Laboratories
Hightstown, New Jersey

and

Hirotoshi Sano

Department of Chemistry
Ochanomizu University
Tokyo, Japan

The lattice-dynamic anisotropy in double-layer molecules of stannous sulfide and dioctyltinoxide was investigated in the temperature range 60–320°K *by observing the Karyagin effect in Mössbauer doublet spectra. It was found that in molecules with a two-dimensional polymeric network the motion of the Mössbauer atom is largest in the direction perpendicular to the plane of the extended layer, and that the lattice-dynamic anisotropy increases at higher temperatures. The "polymer effect" enhances the recoilless fraction by restricting the motion of the molecule in the direction of the polymer bonding. An analysis for the evaluation of absolute values of the recoilless fraction has been outlined. It has long been recognized that many group IVa/VIa binary compounds have characteristic semiconducting properties. In particular,* PbS, PbSe, PbTe, *and* SnTe, *with regular rock-salt structures, are some of the more well-known infrared-sensitive photoconductors. Another prominent set of binary compounds belonging to this group is the series* GeS, GeSe, SnS, *and* SnSe. *This series of isomorphous semiconductors crystallizes in a distorted* NaCl *structure forming a puckered double-layer molecule with a pseudotetragonal array that extends infinitely in two dimensions perpendicular to the crystal c axis. The crystal structures of* GeS, GeSe, SnS, *and* SnSe *indicate that the most probable space group is* $D_{2h}{}^{16}$ *Pcmn and that the stereochemistry of this series is closely analogous to that of black phosphorus. It is shown that stannous selenide is lattice-dynamically and electronically isomorphous with stannous sulfide. The observed lattice-dynamic anisotropy in both* SnS *and* SnSe *would contribute to the anisotropic semiconducting behavior in these materials. It is suggested that a rigid-body treatment as applied to a layer-type structure could be extended to include the motion of atoms on the surface of small particles. A Thirring expansion analysis is used to determine the frequency moment in the high-temperature limit for* SnS. *Analysis of the "effective mass" of electrons and atoms in stannous sulfide is given.*

INTRODUCTION

It has long been recognized that many group IVa/VIa binary compounds have characteristic semiconducting properties. In particular, PbS, PbSe, PbTe, and SnTe, with regular rock-salt structures, are some of the more well-known infrared-sensitive photoconductors.

Another prominent set of binary compounds belonging to this group is the series GeS, GeSe, SnS, and SnSe. This series of isomorphous semiconductors crystallizes in a distorted NaCl structure forming a puckered double-layer molecule with a pseudotetragonal array that extends infinitely in two dimensions perpendicular to the crystal *c* axis. The crystal structures of GeS, GeSe [1], SnS [2], and SnSe [3] indicate that the most probable space group is D_{2h}^{16} *Pcmn* and that the stereochemistry of this series is closely analogous to that of black phosphorus.

In the case of SnS each tin atom is surrounded by six sulfurs and each sulfur atom by six tins, such that the bond lengths can be grouped into a set of three almost orthogonal shorter Sn–S bonds and a set of three longer Sn–S bonds. This type of atomic arrangement will resemble the double-layer configuration of arsenic, antimony, and bismuth and their corresponding sulfur salts. Since a significant (001) cleavage plane exists in SnS, Hofmann [2] concluded that the stronger primary bonding forces are in the plane perpendicular to the crystallographic *c* axis, while the weak secondary bonds are between the adjacent double layers parallel to the *c* axis. It can therefore be expected that this type of bonding will result in lattice-dynamic, electrical, and optical anisotropy.

It has recently been reported that SnS crystals display anisotropic electrical and optical properties [4–7]. For example, it was reported that the hole mobility at 300°K is $\mu_{\|} = 0.18\,\mu_{\perp}$, where $\mu_{\|}$ is the hole mobility parallel to the c axis and $\mu_{\perp}$ is the geometrical mean of the hole mobilities perpendicular to the c axis [5, 7]. From infrared reflectivity measurements [6, 7] it was found that the electron effective mass in the plane perpendicular to the c axis, $m_{\perp}^*$, is given by $m_{\perp}^* = 0.20\,m_0$, whereas parallel to the c axis the effective mass $m_{\|}^* \approx m_0$, where m_0 is the electron mass. Furthermore, it was observed that the electron effective charge e^* and the refractive index n_0 are isotropic with respect to the crystal axes [6]. It was concluded that the lattice vibrations in SnS are isotropic and that the anisotropic hole mobilities are almost entirely due to the anisotropic effective masses of the electrons [6, 7].

We have undertaken this investigation to study the lattice dynamics in SnS and to show that the associated low-frequency lattice vibrational anisotropy cannot be neglected in the discussion of the electrical and optical parameters. Specifically, the angular dependence of the Mössbauer recoilless fraction (the Karyagin effect) has been measured in the temperature range 60–320°K. In addition, we have extended the study to include two-dimensional organometallic tin compounds.

THEORY

The area A_i under the ith resonance line of a normalized Mössbauer-effect energy spectrum is approximately given by [8]

$$A_i = f_s \int_{-\infty}^{\infty} \{1 - \exp[-f_a(\Theta,\Phi; g(T)) N \sigma_i(\Theta,\Phi; E)]\}\, dE \qquad (1)$$

where f_s is the source recoilless fraction, assumed here to be independent of temperature and the direction of emission of the gamma ray; $\sigma_i(\Theta, \Phi; E)$ is the energy-dependent gamma-ray absorption cross-section for the ith nuclear transition with respect to an appropriate frame of reference oriented at spherical angles Θ and Φ relative to the incident photon beam with wavelength $2\pi\lambda$; $f_a(\Theta, \Phi; g(T))$ is the angular-dependent probability of recoilless gamma-ray absorption in the target material containing N atoms capable of resonant absorption;

for example, in the present case N refers to the Sn^{119} concentration in the absorber; and $g(T)$ is a temperature-dependent scalar amplitude function.

Aside from some constant factors, the normalized area will be proportional to two angular- and temperature-dependent parameters,

$$A_i \propto f_a(\Theta, \Phi; g(T)) I_i^0(\Theta, \Phi; h(T)) \tag{2}$$

$I_i^0(\Theta, \Phi; h(T))$ is the total angular-dependent nuclear transition probability for the ith transition. The relevant transition in Sn^{119} is the $M1$–$E2$ transition with ground-state spin $I_g = 1/2+$ and excited-state spin $I_e = 3/2+$. In the presence of a perturbing electric field the doubly degenerate $3/2+$ state splits into two levels with $m = \pm 1/2$ and $m = \pm 3/2$, whereas the ground state remains unsplit, with $m = \pm 1/2$. The temperature-dependent function $h(T)$ has been discussed elsewhere [9].

The perturbation which causes the presently observed Mössbauer doublet spectra is the electrostatic interaction of the atomic electric field gradient (efg) tensor with the nuclear quadrupole moment Q [10, 11]. Letting V_{ij} represent the second derivative of the electrostatic potential due to the total electron density external to a small sphere surrounding the nucleus, the efg tensor is given by

$$\nabla \mathbf{E} = -V_{ij} = -(\partial^2 V / \partial X_i \, \partial X_j) \tag{3}$$

The efg tensor can be diagonalized by choosing a set of mutually orthogonal principal axes (x, y, z) such that the z axis is chosen to coincide with the axis of symmetry and lies along the direction of maximum efg (the major axis), and the x axis is oriented along the minimum efg. Since $\nabla^2 V \equiv 0$ at the nuclear site, we group the efg tensor components so that $|V_{zz}| \geq |V_{yy}| \geq |V_{xx}|$ and define the asymmetry parameter $\eta = (V_{xx} - V_{yy})/V_{zz}$. The parameter η vanishes in the case of axial symmetry about the z axis, and $\eta = 1$ implies $V_{yy} = -V_{zz}$. The magnitude of the quadrupole splitting is

$$\Delta E_Q = \tfrac{1}{2} e V_{zz} Q (1 + \tfrac{1}{3}\eta^2)^{\frac{1}{2}} \tag{4}$$

The orientation of the principal axes of the efg tensor is taken with respect to a set of orthogonal crystallographic axes (a, b, c), where $[\alpha, \beta, \gamma]$ are the Eulerian angles of the efg principal axes relative to (a, b, c). In the case of reduced crystal symmetry, e.g., a triclinic system, a transformation can be performed to orthogonalize the crystal axes. The orthogonal coordinates (a, b, c) are related to the original coordinates (a_0, b_0, c_0) by [12]

$$
\begin{aligned}
a &= a_0 \sin\gamma_0 + c_0(\cos\beta_0 - \cos\alpha_0 \cos\gamma_0)/\sin\gamma_0 \\
b &= b_0 + a_0 \cos\gamma_0 + c_0 \cos\alpha_0 \\
c &= c_0\{\sin^2\alpha_0 - [(\cos\beta_0 - \cos\alpha_0 \cos\gamma_0)/\sin\gamma_0\}^2]^{\frac{1}{2}}
\end{aligned}
$$

The origin of (x, y, z) is taken to be the Sn^{119} nucleus. The direction of the unpolarized incident gamma ray is determined by the polar and azimuthal angles Θ and Φ in the efg principal-axis coordinate system. In the case of a randomly oriented polycrystalline powdered sample the nth crystallite would have its efg principal axes oriented at angles Θ_n and Φ_n with respect to the incident gamma ray.

The radiative transition probabilities for the $(\pm 3/2, 3/2+) \rightarrow (\pm 1/2, 1/2+)$ and the $(\pm 1/2, 3/2+) \rightarrow (\pm 1/2, 1/2+)$ transitions will be denoted by $I_+{}^0$ and $I_-{}^0$, respectively. For the nth crystallite with $\eta = 0$ and $V_{zz} \neq 0$ [13–16], and assuming pure $M1$ radiation, we have

$$I_+{}^0 = K\tfrac{3}{2}(1 + \cos^2\Theta_n) \qquad \text{and} \qquad I_-{}^0 = K\tfrac{3}{2}(\tfrac{5}{3} - \cos^2\Theta_n) \tag{5}$$

For $\eta \neq 0$ and $V_{zz} \neq 0$ the transition probabilities with pure $M1$ radiation for the nth randomly oriented crystallite are [13–16]

$$I_\pm{}^0 = 4\lambda \pm (3\cos^2\Theta_n - 1 + \eta \sin^2\Theta_n \cos 2\Phi_n) \tag{6}$$

where $\lambda = [1 + (\eta^2/3)]^{\frac{1}{2}}$.

Recall that the pertinent transition consists of mixed multipole radiation, but since the ratio of magnetic dipole to electric quadrupole radiation is $M1 > 1 \times 10^3\, E2$ [17], we have justifiably treated the gamma radiation as pure $M1$. A detailed treatment for dipole–quadruple mixtures has been given elsewhere [18, 19].

The recoil-free fraction f_a is related to $\mathbf{k}$, the wave vector of the γ-radiation, and to $\mathbf{u}_r$, the ensemble average of the instantaneous displacement of the rth absorbing nucleus from its mean position. In the harmonic approximation f_a reduces to [20]

$$f_a = |\langle i| \exp(i\mathbf{k}\cdot\mathbf{u}_r)|i\rangle|^2 \tag{7}$$

where $\langle i|$ and $|i\rangle$ refer to the final and initial lattice states, respectively.

For the present analysis we have [21]

$$f_a = \exp[-\langle(\mathbf{k}\cdot\mathbf{u}_r)^2\rangle] \tag{8}$$

For the 23.875-keV Sn^{119m} transition $k^2 = 1.46 \times 10^{18}$ cm^{-2}, and hence $\mathbf{u}_r$ can be determined once an experimental value for f_a is obtained.

A particularly useful analysis of $\mathbf{u}_r$ has been developed in connection with the determination of thermal parameters associated with the molecular X-ray Debye–Waller factor. The lattice-dynamic behavior in molecules can be characterized by the rigid-body [22–24] anisotropic mean-square displacement tensor **U** and the nonrigid-body [25] distortional vibration component $\langle u^2 \rangle_d$. The mean-square displacement of the rth atom in a direction **l** is

$$\langle u_r^2 \rangle = \sum_{ij} U^{ij} l_i l_j + \langle u_r^2 \rangle_d \tag{9}$$

The doubly contravariant nature of **U** and the covariant form of **l** are sufficiently general to make an analysis of noncentrosymmetrical triclinic systems possible. The components of **U**, with respect to a triclinic crystallographic coordinate system, are the contravariant translation tensor **T**, the covariant libration tensor **L**, and the mixed correlation tensor **S** [26], i.e., U^{ij} describes the translation T^{ij} of a rigid body located at **r** and which simultaneously undergoes a rigid-body rotation L_{mr} through an angle ω. The correlation between L_{mr} and T^{ij} is represented by $S_r^{\ m}$ and corresponds to helical type motion of the molecule. The U^{ij} in Eq. (11) can now be written as

$$U^{ij} = T^{ij} + G^{ij,mr} L_{mr} + H_r^{\ ij,m} S_r^{\ m} \tag{10}$$

where the coefficients $G^{ij,mr}$ and $H_r^{\ ij,m}$ are listed in ref. [24].

In order to make the present analysis more tractable, we assume that the crystallographic coordinate axes are orthogonal, that the inertial frame coincides with the crystal axes, that the molecule is centrosymmetrical, that the mass center coincides with the origin of the Sn^{119} efg coordinate system, and that the correlation between **T** and **L** is negligible. These rather stringent conditions are, admittedly, rarely realizable in real cases. However, these simplifications make the distinction between contravariance and covariance no longer necessary, and Eq. (9) reduces to

$$\langle u_r^2 \rangle = \sum_{ij} U_{ij} l_i l_j + \langle u_r^2 \rangle_d$$

and

$$\langle u_r^2 \rangle = \sum_{ij} \{T_{ij} l_i l_j + L_{ij} (\mathbf{l} \Lambda \mathbf{r})_i (\mathbf{l} \Lambda \mathbf{r})_j\} + \langle u_r^2 \rangle_d \tag{11}$$

In order to relate the angular dependence of $\langle u_r^2 \rangle$ with the direction of the incident gamma ray, it is necessary to transform $\langle u_r^2 \rangle$ from the crystal-axis coordinate system to the efg principal axes, where $\langle u_r^2 \rangle$ in the new reference frame will be denoted by $\langle u_s^2 \rangle$. The appropriate transformation is

$$\langle u_s^2 \rangle = \sum_r \mathscr{D}^{(2)}(\{\alpha, \beta, \gamma\})_{sr} \langle u_r^2 \rangle \tag{12}$$

where the matrix $\mathscr{D}^{(2)}(\alpha, \beta, \gamma)$ is the irreducible representation of the rotation group [27]. It is assumed here that both U_{ij} and $\langle u_r^2 \rangle_d$ transform in a similar fashion, although this would probably not be true for most physical situations.

The relation between the various coordinate systems is indicated in Fig. 1, where $\langle u_r^2 \rangle_x$, $\langle u_r^2 \rangle_y$, and $\langle u_r^2 \rangle_z$ represent the principal-axis components of the vibrational ellipsoid described by Eq. (11) and which is graphically shown in Fig. 2.

The angular dependence of f_a with respect to the polar angles Θ and Φ is implicit in $\langle u_s^2 \rangle$. We are now in a position to discuss the ratio of the areas associated with each component of the electric-quadrupole doublet. For a "thin" absorber the intensity ratio of the doublet spectra is equivalent to the corresponding area ratio [28].

When $\eta = 0$ and $V_{zz} > 0$, i.e., where the efg tensor is negative, and at a constant temperature, Eq. (2) in the ratio form becomes

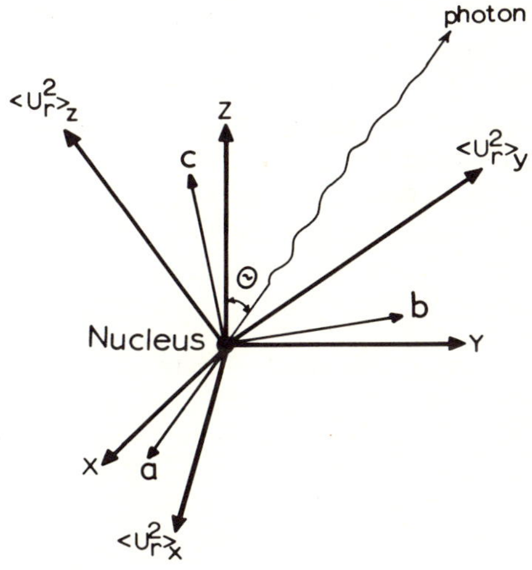

Fig. 1. The spatial relations among the coordinate systems of the efg tensor, the mean-square displacement tensor, and the crystallographic axes.

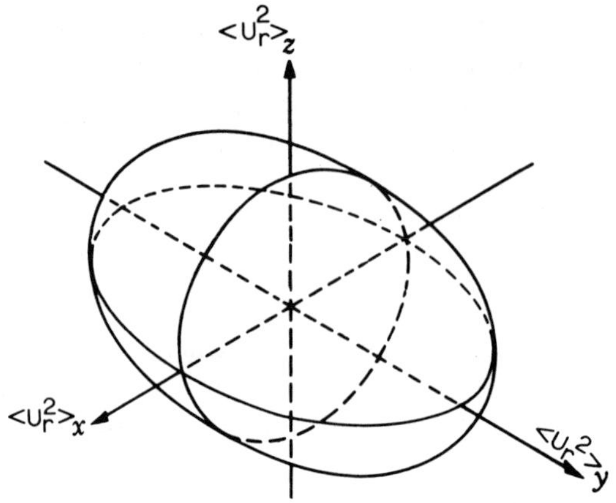

Fig. 2. The principal-axis components of the vibrational ellipsoid.

$$\frac{A_+}{A_-} = \frac{I_2}{I_1} = \frac{\int_0^\pi (1 + \cos^2\Theta) f_a(\Theta) \sin\Theta \, d\Theta}{\int_0^\pi (\frac{5}{3} - \cos^2\Theta) f_a(\Theta) \sin\Theta \, d\Theta} \tag{13}$$

A_+ and A_- or I_2 and I_1 refer to the line components associated with the gamma-ray resonance energies $E_0 + \delta$ and $E_0 - \delta$, respectively, where E_0 is the gamma-ray energy at the centroid of the spectrum. If the lattice dynamic behavior of $\langle u_s^2 \rangle$ is isotropic, i.e., when $\langle u_s^2 \rangle_x = \langle u_s^2 \rangle_y = \langle u_s^2 \rangle_z$ where Eq. (11) describes a vibrational sphere, then f_a is a constant at a given temperature, and upon integration Eq. (13) reduces to unity. More generally, $\langle u_s^2 \rangle$ is anisotropic even at low temperatures, and hence an angular dependence in f_a is expected. In this case $I_2/I_1 \neq 1$ and asymmetries in the quadrupole doublet line intensity for polycrystalline powders will be observed, i.e., the Karyagin effect [14–16].

EXPERIMENTAL

The experimental details have been reported elsewhere [28]. The recoilless fractions were evaluated by extending the method suggested by Lang [8] to the present investigation. A suitable expression for σ_i appearing in Eq. (1) is

$$\sigma_i(E) = (\pi/2)\sigma_0 \Gamma \sum_{i=1}^{2} b_i k_i(E) \tag{14}$$

where b_i is the reduced transition probability for the ith transition and $k_i(E)$ is the energy distribution of the Mössbauer γ-rays. Under ideal conditions $k_i(E)$ represents a Lorentzian distribution, whereas an excessively broadened line is best described by a Gaussian energy profile. The normalization $\int_0^\infty \sum_i b_i k_i(E)\, dE = 1$ applies. The absorption cross section at resonance σ_0 is given by

$$\sigma_0 = 2\pi\lambda\!\!\!^{-}{}^2 \frac{(2I_e + 1)}{(2I_g + 1)}\left(\frac{\Gamma}{\Gamma_a}\right)\frac{1}{(1 + \alpha)} \tag{15}$$

where Γ is the total natural linewidth of the Sn^{119} $3/2^+$ level; Γ_a is the effective absorber linewidth (FWHM); α is the total internal conversion coefficient for the $(3/2^+) \rightarrow (1/2^+)$ transition and has been reported [16] to be 5.12 ± 0.14; it is not a constant, but varies from one compound to another due to changes in electron densities $\sum \psi_{nlm}{}^2(r)$ at the nucleus associated with different materials. Similarly, a temperature and pressure dependence will influence α due to the variation in $\sum \psi_{nlm}{}^2(r)$. Generally, if the isomer shift or quadrupole splitting shows some temperature or pressure dependence, then α will vary correspondingly due to the functional dependence on $\sum \psi_{nlm}{}^2(r)$. Hence σ_0 is not a constant nuclear parameter and varies according to α. This conclusion follows from the assumption that the electronic and nuclear wave functions are well-behaved in any pertinent neighborhood and the approximate relation holds

$$\alpha = C \sum_{L_i} \sum_{L_f} |\langle \psi_{f,ne} | \mathscr{H}_A | \psi_{i,ne} | \rangle^2 \tag{16}$$

where C is a constant and represents the density of final electronic and nuclear states, $\langle \psi_{f,ne}|$ and $|\psi_{i,ne}\rangle$ are the final and initial nuclear and electronic wave functions, respectively, and $\mathscr{H}_A$ is the interaction Hamiltonian between the nuclear and the atomic systems and contains the multipole matrix element. The summation is taken over the angular-momentum quantum numbers of the multipole radiation.

The effective absorber thickness for the ith line is defined as

$$t_i = (\pi/2)\sigma_0 \Gamma b_i N f_a \tag{17}$$

Equation (1) can be simplified to read

$$A_i = f_s \int_{-\infty}^{\infty} \{1 - \exp[-t_i k_i(E)]\}\, dE \tag{18}$$

Upon integration we have for a Lorentzian-shaped spectrum

$$A_i = (\pi/2)f_s\Gamma L(t_i) \tag{19}$$

and for a non-Lorentzian

$$A_i = f_s\Gamma' C_\delta M_\delta(t_i) = f_s\kappa M_\delta(t_i) \tag{20}$$

where $L(t_i)$ and $M_\delta(t_i)$ are saturation functions directly related to t_i and are tabulated in ref. [8] for $t_i \leq 6$. Values for $t_i > 6$ have been listed elsewhere [30].

The κ and δ are determined from either $\lim_{N\to 0^+}\Gamma_e$ or $\lim_{f_a\to 0^+}\Gamma_e$ where [21] the total experimental linewidth Γ_e is given by [31]

$$\Gamma_e = (\Gamma_a + \Gamma_s) + 0.27\,\Gamma N\sigma_0 f_a \tag{21}$$

where Γ_s is the effective source linewidth. Both limits converge to $(\Gamma_a + \Gamma_s)$. Evaluation of $\lim_{N\to 0^+}\Gamma_e$ yields a linear relation where the intercept with $N \to 0$ gives $(\Gamma_a + \Gamma_s)$ directly. The behavior of Γ_e with respect to f_a involves a lattice-dynamic parameter,

$$\Gamma_e = (\Gamma_a + \Gamma_s) + 0.27\,\Gamma N\sigma_0(\exp - [\langle(\mathbf{k}\cdot\mathbf{u}_r)^2\rangle]) \tag{22}$$

It is therefore difficult to estimate this convergence experimentally because it requires that $T \to \epsilon$, where ϵ is a least upper bound of the set of all temperatures associated with the f_a measurements. For this reason we generally found that $\{\lim_{f_a\to 0^+}\Gamma_e\} \geq \{\lim_{N\to 0^+}\Gamma_e\}$.

In order to avoid the complication of line broadening due to unresolvable electric-quadrupole interaction it is recommended here that an absorber material with large ΔE_Q be employed to determine $(\Gamma_a + \Gamma_s)$, e.g., $(CH_3)_2SnCl_2\cdot$2,2′-bipyridyl, or $(C_4H_9)_2SnSO_4$, with $\Delta E_Q = 4.09 \pm 0.09$ mm/sec and $\Delta E_Q = 4.90 \pm 0.09$ mm/sec, respectively. In the absence of relaxation effects and magnetic interactions the well-resolved doublet lines will not include additional broadening due to extranuclear inhomogeneities. This conclusion follows at once from Kramer's degeneracy theorem.

The source recoilless fraction was evaluated from $f_s = A_i/\kappa M_\delta(t_i)$. The experimental area can be approximated by $A_i = (1/2)(c/E_0)\pi\Gamma_e I_i$, where I_i is the line intensity, defined by $I = (N_\infty - N)/N_\infty$, with N_∞ the counting rate at infinite velocity and N the counting rate at the resonance maximum; and c is the vacuum speed of light. The absorber material was white metallic β-Sn, for which absolute recoilless fractions had been reported for various temperatures [32]. The inherent quadrupole splitting in β-Sn ($\Delta E_Q \approx 0.25$ mm/sec) was incorporated in κ. Assuming $b = 1$ for the apparent singlet spectra in β-Sn, t in Eq. (17), and hence $M_\delta(t_i)$, can be calculated for a given foil thickness and temperature. We have assumed that $f_s = f_a$ if the source and absorber materials are chemically and physically identical. It should be noted that the present f_s value does not necessarily correspond to an absolute value of the source recoilless fraction. Instead, f_s here represents an operational quantity which reflects the present experimental conditions. Since this quantity contains all the experimental uncertainties, it permits the calculation of absolute values of f_a.

Absolute values for the absorber recoilless fraction can be obtained by first evaluating $M_\delta(t_i) = A_i/f_s\kappa$ and then finding the corresponding t_i. We assumed here that for a singlet $b = 1$ and for a doublet $b_1 = b_2 = 1/2$, such that $\sum b_i = 1$ and $\sum b_i f_i = f_a$. Thus, we obtain $f_a = 2t/\pi\Gamma\sigma_0 N$ for a singlet and $f_a = (\pi\Gamma\sigma_0 N)^{-1}(2t_1 + 2t_2)$ for a doublet Mössbauer spectrum. The assumption that $f_a = (f_1 + f_2)/2$ for a doublet spectrum displaying the Karyagin effect is not strictly true. In this case a more realistic relation would be $f_a = (mf_1 + nf_2)/(m + n)$, where m and n are the appropriate weighting factors.

DISCUSSION AND RESULTS

The temperature dependence of f_a for stannous sulfide and di-octyltin oxide is shown in Fig. 3, where $\log_{10} f_a$ has been plotted

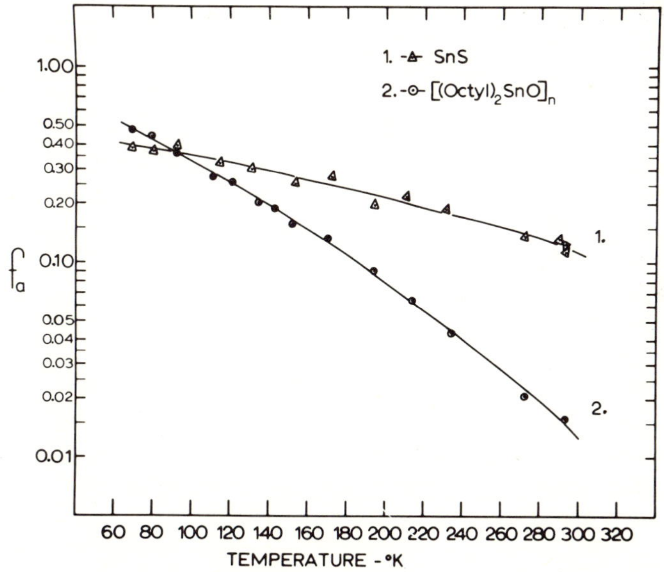

Fig. 3. The temperature dependence of f_a ($\log_{10}$ scale) for SnS (curve 1), and $(Octyl)_2SnO$ (curve 2).

against the temperature T in degrees Kelvin. In the range $60 < T < 300$ the following relation holds:

$$\log_{10} f_a = a_0 + a_1 T + a_2 T^2 \tag{23}$$

The coefficients a_0, a_1, and a_2 are listed in Table I. From Eq. (8) we have

$$f_a = \exp(-2W) \tag{24}$$

where we define

$$2W = k^2 \langle u_r^2 \rangle = \sum_{ij} U_{ij} k_i k_j + k_{ij}^2 \langle u_r^2 \rangle_d \tag{24'}$$

Table I. The Coefficients of the Least-Squares-Fit Polynomial Describing the Temperature Dependence of f_a

Compound	a_0	a_1	a_2
SnS	−0.3051	−0.0011756	−0.000003027
$[(\mathrm{Octyl})_2\mathrm{SnO}]_n$	−0.01728	−0.003805	−0.000007879
$\mathrm{Sn(TDT)_2}$	+0.002146	−0.00443	—

Using Eq. (11) we have

$$2W = \sum_{ij} \{T_{ij} k_i k_j + L_{ij} (\mathbf{k\Lambda r})_i (\mathbf{k\Lambda r})_j\} + k_{ij}^2 \langle u_r^2 \rangle_d \tag{25}$$

The presently reported mean-square displacements of the absorbing nucleus were calculated from $\langle u_r^2 \rangle = 2Wk^{-2}$. The temperature dependence of $\langle u_r^2 \rangle$ for stannous sulfide and di-octyltin oxide is shown in Fig. 4. In the temperature interval $60 < T < 300$, $\langle u_r^2 \rangle$ fits the equation

$$\langle u_r^2 \rangle = [b_0 + b_1 T + b_2 T^2] \times 10^{-18}\ \mathrm{cm}^2 \tag{26}$$

The coefficients b_0, b_1, and b_2 are listed in Table II. For the sake of completeness we have included the temperature dependence of f_a and $\langle u_r^2 \rangle$ for bis-3,4-toluenedithiol tin ($\mathrm{Sn(TDT)_2}$) in the range $60 < T < 330$.

The complete interpretation of $\langle u_r^2 \rangle$ for a molecular crystal is hardly possible at this time. However, a relatively simple expression

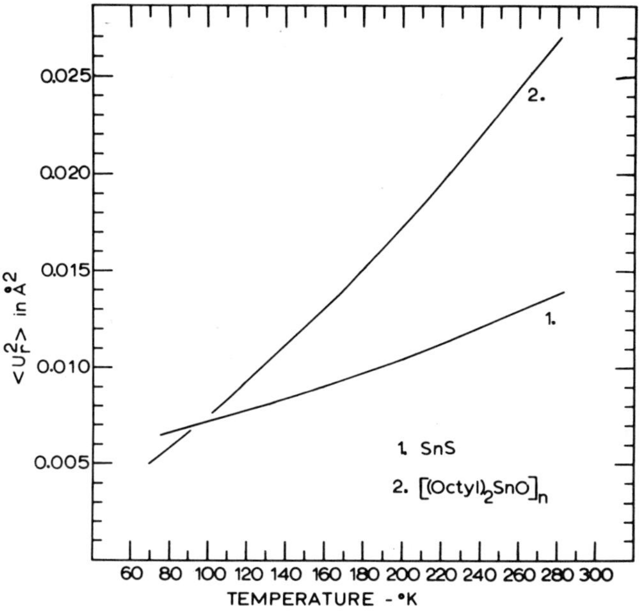

Fig. 4. The temperature dependence of the mean-square displacement in units of (angstroms)^2 for SnS (curve 1) and $(Octyl)_2SnO$ (curve 2).

Table II. The Coefficients of the Least-Squares-Fit Polynomial Describing the Temperature Dependence of $\langle u_r^2 \rangle$

Compound	b_0	b_1	b_2
SnS	0.4838	0.001864	0.0000048
$[(Octyl)_2SnO]_n$	0.02741	0.006035	0.0000125
$Sn(TDT)_2$	−0.0034045	0.007035	—

can be obtained for an ideal cubic, monatomic, continuous, harmonic crystal composed of R unit cells with N_0 atoms of mass M as

$$\langle u_r^2 \rangle = \sum_{j=1}^{3n} \frac{\epsilon_j(\mathbf{q})}{RM\omega_j^2(\mathbf{q})} |\hat{\mathbf{Q}}_j(r,\mathbf{q})|^2 \tag{27}$$

There are $3n$ lattice waves with circular frequency $\omega_j(\mathbf{q})$ for any of the R wave vectors $\mathbf{q}$. The indices are $j = 1 \ldots 3n$. The $\hat{\mathbf{Q}}_j(r, \mathbf{q})$ is the polarization vector corresponding to the rth atom with wave vector $\mathbf{q}$.

The mean energy $\epsilon_j(\mathbf{q})$ per vibrational mode for a harmonic oscillator is

$$\epsilon_j(\mathbf{q}) = \hbar\omega_j(\mathbf{q})\{\tfrac{1}{2} + [\exp(\hbar\omega_j(\mathbf{q})/k_B T) - 1]^{-1}\} \qquad (28)$$

where k_B is the Boltzmann constant. For sufficiently large R Eq. (27) can be written in integral form [33]

$$\langle u_r^2 \rangle = (1/RM) \int E(\omega) g(\omega) \omega^{-2} |\hat{Q}(r,\omega)|^2 \, d\omega \qquad (29)$$

where $E(\omega) = \sum_{jq} \epsilon_j(\mathbf{q})$ and $\hat{Q}(r, \omega)$ is the value of $\hat{\mathbf{Q}}_j(r, \mathbf{q})$ for the mode whose frequency is $\omega_j(\mathbf{q})$. Here $g(\omega)\, d\omega$ is the frequency distribution of the normal modes. It should be noted that the result in Eq. (27) is based on a potential energy function φ whose second-order component $\varphi(2)$ is harmonic and is given by [33]

$$\varphi(2) = \tfrac{1}{2} \sum_{r_1} \sum_{r_2} \varphi_{x_i x_j}(r_1, r_2) u_{x_i}(r_1) u_{x_j}(r_2) \qquad (30)$$

The force constant is then defined by

$$\varphi_{x_i x_j}(r_1 r_2) = \partial^2 \varphi / \partial x_{i r_1} \partial x_{j r_2} \qquad (31)$$

This type of analysis has been extended to include simple cubic diatomic systems [34]. A similar treatment for molecular crystals is, however, not entirely valid.

A molecular crystal is usually not continuous, but instead suffers discontinuities at each molecular or unit-cell boundary. Although a regular short-range crystal structure may be present, the lattice dynamics of a molecular system may not be periodic in the reciprocal lattice, and hence the dispersion relations of $\hat{\mathbf{Q}}_j(r, \mathbf{q})$ are no longer restricted along the axes of highest symmetry, i.e., threefold degeneracy. The phonon dispersion takes on an n-fold degeneracy. In addition, a unique potential energy function must be replaced by one which includes covalent and/or ionic intramolecular forces and van der Waals or Lennard-Jones (6–12) intermolecular interaction potentials. When the molecular units are bonded to form a long-range polymeric lattice network it becomes necessary to describe the intermolecular bonding in terms of covalent or ionic potentials, and in this case some of the intra- and intermolecular forces become similar.

The primary contribution to $\langle u_r^2 \rangle$ in a molecular crystal originates from lattice modes where the molecules move almost as rigid units. The high-frequency distortional modes are almost independent of $\mathbf{q}$. Since most lattice modes are composed of mixtures of librational and translational modes for any $\omega_j(\mathbf{q})$, the potential energy is approximately [33]

$$\varphi(2) = \sum_{r_1 r_2} \sum_{x_i x_j} \{\varphi_{x_i x_j}{}^{(1)}(r_1 r_2) u_{x_i}(r_1) u_{x_j}(r_2) + a\varphi_{x_i x_j}{}^{(2)}(r_1 r_2) u_{x_i}(r_1)\theta_{x_j}(r_2)$$
$$+ a\varphi_{x_i x_j}{}^{(2)}(r_1 r_2) u_{x_j}(r_2)\theta_{x_i}(r_1) + a^2\varphi_{x_i x_j}{}^{(3)}(r_1 r_2)\theta_{x_i}(r_1)\theta_{x_j}(r_2)\} \quad (32)$$

where $\varphi_{x_i x_j}{}^{(1)}$, $\varphi_{x_i x_j}{}^{(2)}$, and $\varphi_{x_i x_j}{}^{(3)}$ are the force constants; and $u(r)$ and $\theta(r)$ are the translational and librational displacements, respectively. In the case of a centrosymmetrical molecule we have for the translational displacement

$$\langle u_T{}^2 \rangle_x = (c_1/M_1{}^*) \int E(\omega) g_T(\omega) \omega^{-2}\, d\omega \quad (33)$$

and for the librational displacement

$$\langle \theta_L{}^2 \rangle_\alpha = (c_2/K^2 M_2{}^*) \int E(\omega) g_L(\omega) \omega^{-2}\, d\omega \quad (34)$$

where $g_T(\omega)$ and $g_L(\omega)$ are the translational and librational frequency distributions, respectively,

$$g(\omega) = g_T(\omega) + g_L(\omega) \quad (35)$$

K is the molecular radius of gyration, and M^* is the effective mass of the molecule.

The atomic mass, and hence the molecular weight, is generally obtained from mass-spectroscopic data in terms of an *inertial mass* or from nuclear-reaction Q values that yield an energy–equivalent mass relation. The atomic *inertial mass* M represents approximately a *free atom mass*, and unless the complete equation of motion for a molecular crystal is known, the exact meaning of M in a solid is not clear. The incorporation of M in the Mössbauer recoilless fraction may therefore lead to an artificial interpretation of $2W$ even for the simplest crystalline systems. Alternatively, however, an atom bound in a molecular crystal or experiencing some external forces could perhaps be represented by an *effective-mass tensor* M^*, which is somewhat reminiscent of the *energy-mass* mentioned above. As we shall see below, M^* includes the influence of lattice forces on an atomic unit, and hence presupposes a knowledge of the exact energy-phonon wave dispersion relationship. According to the duality of mass, the *effective-mass tensor* is equivalent to the *inertial mass* in the case of an ideal crystal, where all the forces and the equation of motion are known.

A lattice-dynamic expression for M^* can be obtained by letting $\mathbf{P}$ represent the crystal momentum as $\mathbf{P} = \hbar\mathbf{q}$. The group velocity of the phonon is given by the following vector field in $\mathbf{P}$ space

$$\mathbf{v} = \nabla_p E \tag{36}$$

where for a harmonic case $E = \sum_{jq}\epsilon_j(\mathbf{q})$, and $\epsilon_j(\mathbf{q})$ is given by Eq. (28); $\mathbf{v}$ is also the average instantaneous velocity. Similarly, in $\mathbf{q}$ space we have

$$\mathbf{v} = \hbar^{-1}\nabla_q E \tag{37}$$

A force $\mathbf{F}$ acting on a mass center will give rise to a time rate of change in energy, and hence

$$\dot{E} = \nabla_p E \cdot \dot{\mathbf{P}} = \nabla_p E \cdot \mathbf{F} = \mathbf{v} \cdot \mathbf{F} \tag{38}$$

The average acceleration is given by

$$\dot{\mathbf{v}} = \nabla_p \dot{E} = \nabla_p(\nabla_p E \cdot \mathbf{F}) = \mathbf{O}E \cdot \mathbf{F} \tag{39}$$

where $\mathbf{O}$ is a second-rank tensor operator. The equation of motion of the rigid molecule (and averaging over all internal modes) is

$$M_1{}^* \, \partial^2 u_{x_i}(r_1)/\partial t^2 = -\partial\varphi(2)/\partial u_{x_i}(r_1) \tag{40a}$$

and

$$M_2{}^* K^2 \, \partial^2 \theta_{x_i}(r_1)/\partial t^2 = -\partial\varphi(2)/\partial \theta_{x_i}(r_1) \tag{40b}$$

If we now compare Eq. (39) with the equation of motion, an expression for M^* is obtained as

$$1/M^* = \mathbf{O}E = \partial^2 E/\partial P_i \, \partial P_j = (1/\hbar^2)(\partial^2 E/\partial q_i \, \partial q_j) \tag{41}$$

According to Eq. (41), the atomic mass or the molecular mass in a solid is no longer a scalar quantity, but now becomes a tensor representation. For an isotropic crystalline solid $(M^*)^{-1}$ describes an *effective mass sphere*, whereas for an anisotropic solid the molecular mass is an *effective mass ellipsoid*. Equation (41) can be transformed to yield the principal-axis components of the spheroid as $M_{xx}{}^*$, $M_{yy}{}^*$, and $M_{zz}{}^*$. Note that $M_{ii}{}^*$ will generally depend upon the type of motion under discussion. Thus, $M_{ii}{}^*$ will usually be different for a molecule executing rigid-body motion and nonrigid distortional atomic motion, because the potential energies and the dispersion relationships associated with intra- and intermolecular motion will differ considerably. This treatment could possibly explain the earlier experimental observation that almost no systematic correlation was found between the recoil-free fraction and the molecular mass in organometallic compounds [36].

Although it is possible to describe $\langle u_r^2 \rangle$ in terms of **U** and $\langle u_r^2 \rangle_d$, i.e., $\langle u_T^2 \rangle_x$ and $\langle \theta_L^2 \rangle_\alpha$, the Mössbauer scattering effect can only measure the time-average thermal ensemble $\langle u_r^2 \rangle$ directly. The time-averaging occurs over the lifetime of the excited state, and is therefore considered a "slow process" compared to the "fast process" found in X-ray scattering [37]. Nevertheless, since it was observed earlier that $\langle u_r^2 \rangle$ is strongly dependent upon the nature of the polymeric lattice structure, which directly involves the bonding of the Mössbauer atom [36], it can be expected that a detailed study of such a "polymer effect" may elucidate the behavior of $\langle u_r^2 \rangle$. Thus, it was shown that for linear chain molecules and layer-type structures $\langle u_r^2 \rangle$ is smallest along the chain axis and in the plane of the layer. We can develop the present analysis further according to the theory due to Karyagin.

Since the ratio of the Mössbauer line intensities is dependent upon the angular dependence of the recoil-free fraction, it is possible to evaluate the anisotropy in $\langle u_r^2 \rangle$. If it is assumed that the pertinent part of the mean-square displacement is translational with respect to Θ and Φ, i.e., $\langle u_s^2 \rangle = \sum_{ij} T_{ij} k_i k_j$ and that the off-diagonal terms are zero, then

$$\langle u_s^2 \rangle = \langle u_s^2 \rangle_x \sin^2\Theta \cos^2\Phi + \langle u_s^2 \rangle_y \sin^2\Theta \sin^2\Phi + \langle u_s^2 \rangle_z \cos^2\Theta \tag{42}$$

where the directions x, y, and z coincide with the transformed principal axes of elasticity of the crystal. For an axially symmetrical lattice-dynamic structure $\langle u_s^2 \rangle_x = \langle u_s^2 \rangle_y = \langle u_s^2 \rangle_p$, and Eq. (42) reduces to

$$\langle u_s^2 \rangle = \langle u_s^2 \rangle_p \sin^2\Theta + \langle u_s^2 \rangle_z \cos^2\Theta \tag{43}$$

In the case of SnS the space group D_{2h}^{16} implies almost complete centrosymmetry, and we can calculate the lattice anisotropy to a first approximation. The experimental intensity ratios I_2/I_1 at various temperatures for SnS, $[(\text{Octyl})_2\text{SnO}]_n$, and Sn(TDT)_2 are shown in Fig. 5. The temperature-dependent behavior of I_2/I_1 is given by the equation

$$I_2/I_1 = c_0 + c_1 T + c_2 T^2 + c_3 T^3 \tag{44}$$

where the coefficients are listed in Table III.

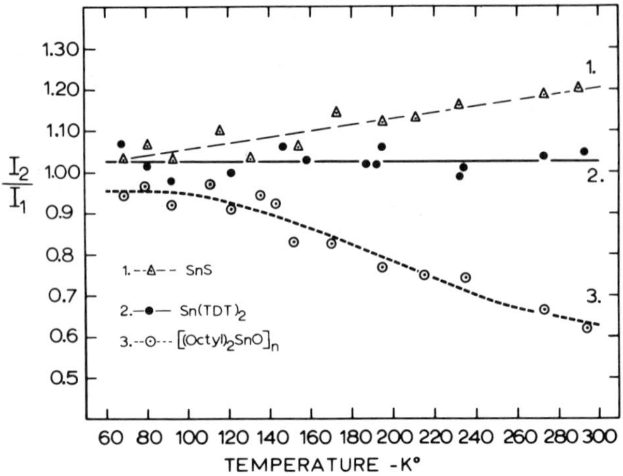

Fig. 5. The temperature dependence of the line intensity ratio, I_2/I_1, for SnS (curve 1), $Sn(TDT)_2$ (curve 2), and $(Octyl)_2SnO$ (curve 3).

Table III. The Coefficients of the Least-Squares-Fit Polynomial Describing the Temperature Dependence of I_2/I_1

Compound	c_0	c_1	c_2	c_3
SnS	0.97899	0.7523×10^{-2}	—	—
$[(Octyl)_2SnO]_n$	0.81992	0.371543×10^{-2}	-0.2953×10^{-4}	0.05019×10^{-6}
$Sn(TDT)_2$	1.02685	—	—	—

Assuming that $V_{zz} < 0$ and $\eta \approx 0$ for SnS [42], the anisotropy can be evaluated from I_1/I_2. The resultant equation for SnS becomes

$$I_1/I_2 = 1.00963 - 6.0996 \times 10^{-4} T \tag{45}$$

On the assumption that $\langle u_r^2 \rangle_y - \langle u_r^2 \rangle_x = 0.077 \times 10^{-18}$ cm^2 for SnS over the entire temperature range, we can now use the machine calculations reported by Travis [43] to evaluate the temperature-dependent lattice anisotropy in terms of $(\langle u_r^2 \rangle_z - \langle u_r^2 \rangle_y)$. The results are shown graphically in Fig. 6. The equation governing this behavior is

$$(\langle u_r^2 \rangle_z - \langle u_r^2 \rangle_y) = \{-0.26264 + 6.549 \times 10^{-3} T - 2.033 \times 10^{-5} T^2 + 4.935 \times 10^{-8} T^3\} \times 10^{-18}\ \text{cm}^2 \tag{46}$$

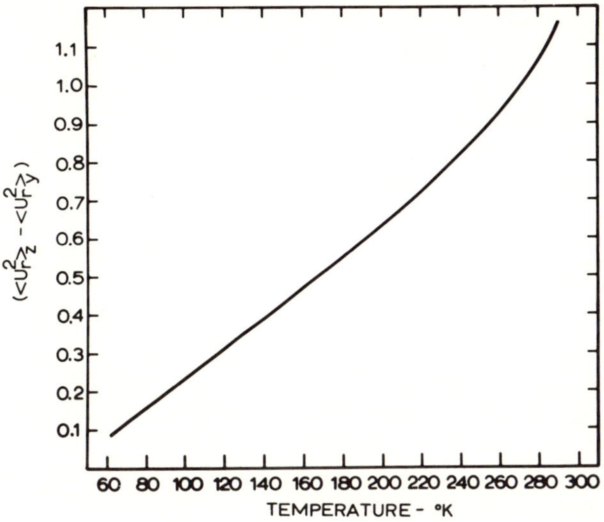

Fig. 6. The temperature dependence of the anisotropic mean-square displacement factor $(\langle u_r^2\rangle_z - \langle u_r^2\rangle_y)$ for SnS.

It can readily be seen that the lattice anisotropy increases at higher temperature, where the electron–phonon scattering may become more prominent.

The tin in dioctyl tin oxide is pentacoordinated with three nearest-neighbor oxygen atoms to form a two-dimensional polymeric CdI_2-type layer 1 lattice [44] with a well-ordered intermolecular arrangement [45]. From Fig. 5 and Eq. (44), it can be observed that at low temperatures the polymeric layer lattice can be approximated by a vibrational sphere, i.e., $\langle u_r^2\rangle_x \approx \langle u_r^2\rangle_y \approx \langle u_r^2\rangle_z$. At high temperatures a significant anisotropic thermal behavior is observed. The eccentricity of the vibrational ellipsoid is maximized with $(\langle u_r^2\rangle_x \approx \langle u_r^2\rangle_y) \neq (\langle u_r^2\rangle_z)$.

The Thirring expansion has been applied to Eq. (27) by Feldman and Horton [46] to obtain the frequency moments for $\langle u_r^2\rangle$. We have then for the first three terms

$$\langle u_r^2\rangle = \frac{k_B T}{M\omega_j^2(-2)} + \frac{\hbar^2}{12\, Mk_B T} - \frac{\hbar^4}{720\, Mk_B^3 T^3}\omega_j^2(2) \qquad (47)$$

where $\omega_j^2(-2)$ and $\omega_j^2(2)$ are the squares of the even frequency moments. If M is taken to be the mass of the Sn^{119} atom, $\langle u_r^2\rangle$ is

$$\langle u_r^2 \rangle = \frac{6.97 \times 10^2}{\omega^2(-2)} T + 3.38 \times 10^{-18} T^{-1} - 3.28 \times 10^{-42} T^{-3} \omega^2(2) \tag{48}$$

$\omega^2(-2)$ can be evaluated from the equation $y = mT^4 + b$, where we defined $y = (\langle u_r^2 \rangle T^3 - 3.38 \times 10^{-18} T^2)$, $m = 6.97 \times 10^5/\omega^2(-2)$, and $b = -3.28 \times 10^{-42}\omega^2(2)$. The curve of y *versus* T^4 gives $\omega^2(-2)$ directly from the slope. The linearity of such an equation at high temperature is a rather severe test on the constancy of $\omega^2(-2)$, and hence on the harmonicity of the crystal, since the temperature goes as the fourth power of T. We found that for SnS $\omega^2(-2)$ was extraordinarily constant between 150°K and 280°K, and $\omega^2(-2) = 1.45 \times 10^{26}$ sec^{-2}. We can define a lattice-dynamic parameter $\theta_\infty = (\hbar/k_B)[(5/3)^{\frac{1}{2}}\omega(-2)]$ and find $\theta_\infty = 116.2$°K for SnS. The quality θ_∞ is frequently associated with the Debye–Waller factor on the assumption that SnS behaves according to the Debye theory. We have seen above that the analysis of the Karyagin effect indicates that SnS is anisotropic even at high temperatures, and hence a Debye theory is not applicable here.

A similar treatment for dioctyl tin oxide yields $\omega(-2) = 8.41 \times 10^{12}$ sec^{-1} and $\theta_\infty = 82.8$°K. For the sake of comparison we have evaluated Hohenemser's [32] data for metallic β-Sn and found $\omega(-2) = 9.68 \times 10^{12}$ sec^{-1} and $\theta_\infty = 95.35$°K. The pertinent temperature range is 200–280°K.

In the case of $Sn(TDT)_2$ the relatively smaller mean-square displacement is due to the polymeric network in this crystal [36]. However a negligible Karyagin effect is observed due to the fact that $Sn(TDT)_2$ is an almost isotropic three-dimensional polymer where the sulfur atoms bridge the tin atoms, and hence the motion of the molecule can be approximated by a vibrational sphere. The small deviation from isotropy, i.e., $(I_2/I_1) \neq 1$, can be attributed to characteristic molecular packing in such organotin compounds.

For the sake of completeness we have analyzed stannic sulfide, since it has a CdI_2-type layer structure with space group D_{3d}^3 P_3mI and it is expected that the lattice dynamics are strongly anisotropic in SnS_2. This anisotropy in polycrystalline, powdered SnS_2 cannot, however, be analyzed by the Karyagin effect, due to the absence of resolvable doublets in this case. Some insight into the nature of the lattice anisotropy can, nevertheless, be gained by investigating the angular dependence of the recoilless fraction in single-crystal platelets of SnS_2.

The hyperfine-interaction results for the compounds discussed here are listed in Table IV. The absence of any significant resolvable

Table IV. Mössbauer Hyperfine Interaction Parameters for Various Compounds

Compound	Monopole interaction (isomer shift),* mm/sec ± 0.09	Quadrupole interaction ΔE_Q (quadrupole splitting) at 295°K, mm/sec ±0.09
SnS_2	1.16	—
SnS	3.31	0.80
SnSe	3.31	0.74
$[(Octyl)_2SnO]_n$	0.96	2.05

*At 295°K; with respect to SnO_2.

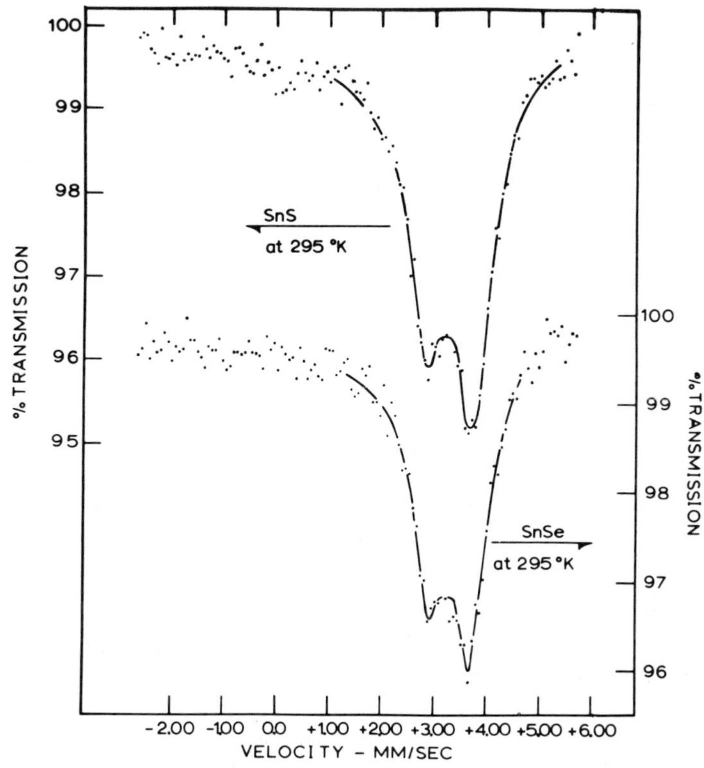

Fig. 7. The Mössbauer spectra for SnS and SnSe at 295°K. The source is $BaSn^{119m}O_3$ at 295°K moving relative to the stationary absorber. The line asymmetry is clearly illustrated.

quadrupole splitting and the isomer shift characteristic of $S^m p^n d^k$ hybridization in tin for SnS_2 agrees with the earlier conclusion [47] that SnS_2 is primarily covalent with octahedral Sp^3d^2 bonds being formed with six nearest-neighbor sulfur atoms. An isomer shift of 3.31 ± 0.09 mm/sec with respect to SnO_2 for both SnS and SnSe reflects the divalency of tin(II) [48] with almost completely populated 5*s* orbitals and hence a high spherical electron density at the tin nucleus. SnS and SnSe have primarily covalent bonding, and the similarity of isomer shifts for SnS and SnSe suggests that electronegativity effects are probably a minor contribution to the isomer shift. The quadrupole splitting of ~ 0.80 mm/sec for SnS and SnSe can be related to the reduced crystal symmetry, where for the three short bonds to tin(II) the angles S—Sn—S are 88° (two) and 96°, and the angles Se—Sn—Se are 89° (two) and 96°. The Mössbauer spectra of SnS and SnSe are shown in Fig. 7. The similarity of the hyperfine interaction and the well-defined Karyagin effect are clearly observable. The present isomer-shift and quadrupole-splitting data for dioctyltinoxide confirm the previously reported notion that the tin–oxygen bond is relatively ionic, with perhaps a minimum of dative-bond character, which presumably gives rise to a pseudoionic lattice structure [45].

REFERENCES

1. S. N. Dutta and G. A. Jeffrey, *Inorg. Chem.* **4**:1363 (1965).
2. W. Hofmann, *Z. Krist.* **92**:161 (1935).
3. A. Okazaki and I. Ueda, *J. Phys. Soc. Japan* **11**:470 (1956).
4. H. Gobrecht and A. Bartschat, *Z. Physik* **149**:511 (1957).
5. W. Albers, C. Haas, and F. vander Maesen, *J. Phys. Chem. Solids* **15**:306 (1960).
6. C. Haas and M. M. G. Corbey, *J. Phys. Chem. Solids* **20**:197 (1961).
7. W. Albers, C. Haas, H. Ober, G. R. Schodder, and J. D. Wasscher, *J. Phys. Chem. Solids* **23**:215 (1962).
8. G. Lang, *Nucl. Instr. Methods* **24**:425 (1963).
9. H. A. Stöckler and H. Sano, *Physics Letters* **25A**:550 (1967).
10. A. Abragam, *The Principles of Nuclear Magnetism* (Clarendon Press, Oxford University, 1961), Chap. VI.
11. H. A. Stöckler, H. Sano, and R. H. Herber, *J. Chem. Phys.* **45**:1182 (1966).
12. D. W. J. Cruickshank *et al.*, *Computing Methods and the Phase Problems in X-Ray Crystal Analysis* (Pergamon Press, New York, 1961), p. 71.
13. P. Zory, *Phys. Rev.* **140**:A1401 (1965).
14. S. V. Karyagin, *Dokl. Akad. Nauk. SSSR* **148**:1102 (1963) [English transl.: *Proc. Acad. Sci. USSR, Phys. Chem. Sect.* **148**:110 (1963)].
15. S. V. Karyagin, *Fiz. Tverd. Tela.* **5**:2128 (1963) [English transl.: *Soviet Physics—Solid State*, **5**:1552 (1964)].
16. S. V. Karyagin, *Fiz. Tverd. Tela* **8**:1739 (1966); **9**:2514 (1967) [English transl.: *Soviet Physics—Solid State* **8**:1387 (1966); **9**:1979 (1967)].

17. J. P. Bocquet, Y. Y. Chu, G. T. Emery, and M. L. Perlman, *Phys. Rev.* **167**:1117 (1968).
18. H. Frauenfelder, D. E. Nagle, R. D. Taylor, D. R. F. Cochran, and W. M. Vissacher, *Phys. Rev.* **126**:1065 (1962).
19. J. T. Dehn, J. G. Marzolf, J. F. Salmon, *Phys. Rev.* **135**:B1307 (1964).
20. L. Van Hove, *Phys. Rev.* **95**:249 (1954).
21. J. Petzold, *Z. Physik* **163**:71 (1961).
22. E. W. J. Cruickshank, *Acta Cryst.* **9**:754 (1956).
23. V. Schomaker and K. N. Trueblood, *Acta Cryst.* **21**:A247 (1966).
24. D. M. Burns, W. G. Ferrier, and J. T. McMullan, *Acta Cryst.* **23**:1098 (1967).
25. K. Lonsdale, *Acta Cryst.* **14**:37 (1961).
26. V. Schomaker and K. N. Trueblood, *Acta Cryst.* **B24**:63 (1968).
27. E. P. Wigner, *Group Theory and Its Application to the Quantum Mechanics of Atomic Spectra* (Academic Press, New York, 1959).
28. H. Stöckler and H. Sano, *Nucl. Instr. Meth.* **44**:103 (1966).
29. H. Stöckler and H. Sano, *Trans. Faraday Soc.* **64**:577 (1968).
30. D. W. Hafemeister and E. B. Shera, *Nucl. Inst. Meth.* **41**:133 (1966).
31. D. A. O'Connor, *Nucl. Inst. Meth.* **24**:425 (1963).
32. C. Hohenemser, *Phys. Rev.* **139A**:185 (1965).
33. W. Cochran, *Rep. Progr. Phys.* **26**:1 (1963).
34. S. S. Jaswal, *J. Chem. Phys.* **46**:4311 (1967).
35. W. Cochran and G. S. Pawley, *Proc. Roy. Soc.* **280A**:1 (1964).
36. H. A. Stöckler, H. Sano, and R. H. Herber, *J. Chem. Phys.* **47**:1567 (1967).
37. R. L. Mössbauer and D. H. Sharp, *Rev. Mod. Phys.* **36**:410 (1964).
38. H. A. Stöckler and H. Sano, *Phys. Rev.* **165**:406 (1968).
39. H. A. Stöckler and H. Sano, in: *Molecular Dynamics and Structure of Solids*, Proceedings 2nd Materials Research Symposium (National Bureau of Standards, US Department of Commerce, Washington, D.C. 1967).
40. H. Sano, M. Aratani, and H. A. Stöckler, *Physics Letters* **26A**:559 (1968).
41. R. W. James, *The Optical Principles of the Diffraction of X-Rays* (Cornell University Press, New York, 1965), Chap. V.
42. A. J. F. Boyle, D. St. P. Bunbury, and C. Edwards, *Proc. Phys. Soc.*, **79**:416 (1962).
43. J. C. Travis, PhD Thesis, The University of Texas, August 1967.
44. D. L. Alleston, A. G. Davies, and M. Hancock, *J. Chem. Soc.* **1964**:5744.
45. W. T. Reichle, *J. Polymer Sci.* **49**:521 (1961).
46. J. L. Feldman and G. K. Horton, *Phys. Rev.* **132**:644 (1963).
47. A. F. Wells, *Structural Inorganic Chemistry* (Oxford University Press, London 1962), 3rd ed., p. 902.
48. R. E. Rundle and D. H. Olson, *Inorg. Chem.* **3**:596 (1964).

Mössbauer Spectroscopy of Inorganic Antimony Compounds

J. G. Stevens* and L. H. Bowen

Chemistry Department
North Carolina State University
Raleigh, North Carolina

In recent years Mössbauer effect spectroscopy (MES) *has been applied to the study of antimony compounds by groups at Argonne National Laboratory* [1, 2] *and Moscow State University* [3]. *This investigation extends the measured compounds to include all the simple oxides, sulfides, and halides of antimony. The use of* MES *for determination of oxidation state, bond character, and structural features is discussed. The shapes of the spectra have been studied in order to obtain additional parameters. The parameters which affect shape include the ratio of the quadrupole moment of the excited state to the ground state of* Sb^{121}, *R; the quadrupole splitting, QS; and the asymmetry factor,* η.

EXPERIMENTAL PROCEDURE

The source was prepared by irradiating Sn^{120} for $\sim$7 months in a thermal neutron flux of $\sim 10^{15}$ n/cm^2/sec. A chemical separation was performed to remove the interfering activities Sb^{125} and Te^{125m}. The final form of the source was $\sim$170 mg of SnO_2, which had an activity of $\sim$150 μCi Sn^{121}. The desired 37.2-keV transition was detected using a xenon–methane proportional tube. The count rate out of the single-channel analyzer (set on the escape peak of the 37.2-keV gamma) was about 1000 cps.

All the compounds were commercially obtained, except for the different crystalline modifications of Sb_2O_3, which were prepared by heat treatment of the commercial compound Sb_2O_3. The crystalline-phase purity of the oxides was checked by X-ray diffraction. All the

*Presently at the Chemistry Department, University of North Carolina, Asheville, North Carolina.

compounds except $SbCl_5$ were powdered, mixed with powdered polyethylene, and placed in a holder having an inside diameter of 1.5 in. The SbF_5, a liquid at room temperature, was frozen in liquid N_2 and then powdered cold. A special liquid holder was used for the $SbCl_5$ sample. The source and absorber were cooled in a liquid-nitrogen cryostat. The absorber temperature was 78–79°K and the source 78–84°K. The source was vibrated by a constant-acceleration drive and the counting data were obtained in the usual manner using a 512-channel analyzer in the time mode.

RESULTS AND DISCUSSION

Spectra

All the spectra, except that of α-Sb_2O_4, are single peaks. The peaks of some spectra are made up of eight lines caused by the Sb^{121} quadrupole splitting (*QS*) interaction [1]. These lines are not resolved because their widths are of the same order of magnitude as the *QS*.

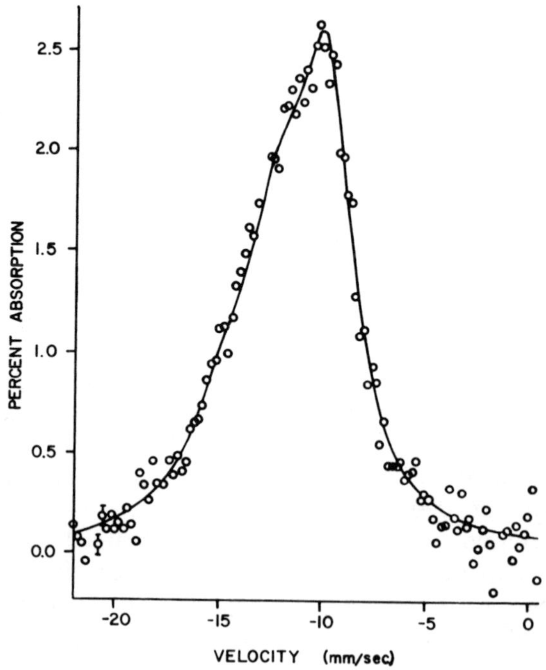

Fig. 1. Mössbauer spectrum of cubic Sb_2O_3.

The result is that the QS affects the shape of the peak, as illustrated by the spectrum of cubic Sb_2O_3 (Fig. 1). When QS occurs the isomer shift (IS) is not at the maximum point in the peak, so QS must be considered in order to obtain accurate values of the IS.

The velocity scale in all the spectra was calibrated from the known energies of the six MHFS lines of the Mössbauer spectrum of an iron foil using a Pd (Co^{57}) source. All the spectra of the antimony compounds were fitted by an iteration technique to several different functions; i.e., single Lorentzian, double Lorentzian, and Sb^{121} quadrupole interaction. The best function was chosen on the basis of the available structural information and the chi-square values (goodness of fit values). The parameters for the chosen functions are reported in Table I.

Table I. Experimental Values of Isomer Shifts (IS) (relative to SnO_2 Source) and Quadrupole Splittings (QS) for Compounds Studied

Compound	IS, mm/sec	e^2qQ, mm/sec
Oxides		
Cubic Sb_2O_3	-11.32 ± 0.12	18.6 ± 0.7
Orthorhombic Sb_2O_3	-11.33 ± 0.07	17.0 ± 0.5
Amorphous Sb_2O_3	-11.35 ± 0.10	18.6 ± 1.0
α-Sb_2O_4	0.61 ± 0.03, -14.36 ± 0.07	-6.1 ± 1.0, 16.4 ± 0.6
Sb_2O_5	1.06 ± 0.02	-4.3 ± 1.1
Sulfides		
Sb_2S_3 (Stibnite)	-14.32 ± 0.11	—
Sb_2S_3 (Amorphous)	-12.74 ± 0.17	—
Sb_2S_5(I)	-12.27 ± 0.11	—
Sb_2S_5(II)	-12.54 ± 0.11	—
Sb_2S_5(III)	-12.11 ± 0.09	—
Halides		
SbF_3	-14.6 ± 0.2	19.6 ± 0.9
$SbCl_3$	-13.8 ± 0.2	12.2 ± 1.7
$SbBr_3$	-13.9 ± 0.2	9.4 ± 1.8
SbI_3	-15.9 ± 0.3	—
$SbCl_5$	-3.12 ± 0.03	-4.4 ± 1.7
SbF_5	2.23 ± 0.12	—
InSb	-8.56 ± 0.08	—

Comparison of QS values obtained in this investigation and those obtained by NQR (Table II) shows that NQR results are several orders of magnitude more precise than the MES results; but, as previously mentioned, QS must not be ignored in analyzing the MES

Table II. Comparison of NQR and MES Results *

Compound	Mössbauer results e^2qQ, MHz	Mössbauer results η	NQR results e^2qQ MHz	NQR results η	Ref.
Cubic Sb_2O_3	548 ± 12	—	554.83	0	[6]
Orthorhombic Sb_2O_3	509 ± 15	0.51 ± 0.14	541.4	0.36	[28]
Amorphous Sb_2O_3	557 ± 30	—	—	—	—
Sb_2O_5	−129 ± 33	—	—	—	—
SbF_3	587 ± 28	—	1073	0.03	[28]
$SbCl_3$	365 ± 51	—	383.66	0.188	[29]
$SbBr_3$	282 ± 54	—	343.95	0.100	[24]
SbI_3	†	—	169.37	0	[24]

*Both at liquid-nitrogen temperatures.
†Value too small to obtain.

results. It should be noted from Table II that there are several compounds for which the MES values for *QS* are the only ones reported. Also, there is a large discrepancy in the MES and NQR results for SbF_3.

Oxides of Antimony

The bond angles and distances [4, 5] around the antimony in cubic Sb_2O_3 are such that the highest-order symmetry element at the antimony site is a unique threefold axis of rotation. This implies that there is *QS*, but that the asymmetry factor $\eta = 0$, which has been shown to be correct by NQR [6]. Using the *QS* from NQR and the Mössbauer spectrum of cubic Sb_2O_3, $R = 1.32 \pm 0.02$ was obtained. This compares with a value of 1.38 ± 0.02 reported by Ruby *et al.* [1].

The bond angles of orthorhombic Sb_2O_3 imply that there is not only quadrupole splitting, but asymmetry. This has also been verified by NQR [6], and, according to the data (Table II), orthorhombic Sb_2O_3 has the largest reported value of η for antimony compounds, as well as a relatively large *QS* value. The Mössbauer spectrum of this compound was chosen, therefore, to determine if η has any effect on the shape of the spectrum. The appropriate energy levels are listed in Table III, and, with the value of $R = 1.32$, were used to obtain the necessary transition energies (Table IV). These equations are valid for $\eta < 0.5$. Figure 2 shows the energy-level diagram for Sb^{121} and the relative positions and intensities of the allowed transitions, as well as

Table III. Electric-Quadrupole Energy Levels for $I = \frac{5}{2}$ and $I = \frac{7}{2}$

I*	M	Energy level
5/2	±1/2	$-\frac{1}{5}e^2Qq\left(1 + \frac{4}{9}\eta^2 - \frac{172}{729}\eta^4 + \frac{15{,}332}{59{,}049}\eta^6 - \frac{1{,}717{,}580}{4{,}782{,}969}\eta^8 + \ldots\right)$
5/2	±3/2	$-\frac{1}{20}e^2Qq\left(1 - \frac{3}{2}\eta^2 + \frac{23}{24}\eta^4 - \frac{449}{432}\eta^6 + \frac{44{,}675}{31{,}104}\eta^8 + \ldots\right)$
5/2	±5/2	$\frac{1}{4}e^2Qq\left(1 + \frac{1}{18}\eta^2 + \frac{17}{5832}\eta^4 - \frac{143}{944{,}784}\eta^6 - \frac{12{,}587}{612{,}220{,}032}\eta^8 + \ldots\right)$
7/2	±1/2	$-\frac{5}{28}e^2Qq\left(1 + \frac{5}{6}\eta^2 - \frac{311}{216}\eta^4 + \frac{20{,}557}{3888}\eta^6 - \frac{726{,}019}{31{,}104}\eta^8 + \ldots\right)$
7/2	±3/2	$-\frac{3}{28}e^2Qq\left(1 - \frac{31}{30}\eta^2 + \frac{21{,}967}{9000}\eta^4 - \frac{3{,}975{,}973}{450{,}000}\eta^6 + \frac{734{,}970{,}487}{18{,}000{,}000}\eta^8 + \ldots\right)$
7/2	±5/2	$\frac{1}{28}e^2Qq\left(1 + \frac{5}{6}\eta^2 + \frac{25}{216}\eta^4 - \frac{275}{3888}\eta^6 - \frac{25}{93{,}312}\eta^8 + \ldots\right)$
7/2	±7/2	$\frac{7}{28}e^2Qq\left(1 + \frac{1}{30}\eta^2 + \frac{29}{27{,}000}\eta^4 + \frac{1241}{12{,}150{,}000}\eta^6 + \frac{2263}{1{,}458{,}000{,}000}\eta^8 + \ldots\right)$

*Values for $I = \frac{5}{2}$ from Bersohn [30]; for $I = \frac{7}{2}$ from Wang [11].

Table IV. Transition Energies for Sb^{121} in Which the Quadrupole Interaction is Present and $\eta \neq 0$ ($R = 1.32$)

Transition $\pm m(I = 7/2) \rightarrow \pm m(I = 5/2)$	Energy	Clebsch–Gordan coefficient
$7/2 \rightarrow 5/2$	$E_1 = IS + (QS/4)(0.320 - 0.012\eta^2)$	21/84
$5/2 \rightarrow 5/2$	$E_2 = IS + (QS/4)(-0.811 + 0.101\eta^2 + 0.019\eta^4)$	6/84
$5/2 \rightarrow 3/2$	$E_3 = IS + (QS/4)(0.389 - 0.143\eta^2 + 0.213\eta^4)$	15/84
$3/2 \rightarrow 5/2$	$E_4 = IS + (QS/4)(-1.566 + 0.579\eta^2 - 1.384\eta^4)$	1/84
$3/2 \rightarrow 3/2$	$E_5 = IS + (QS/4)(-0.3658 + 0.2844\eta^2 - 1.189\eta^4)$	10/84
$3/2 \rightarrow 1/2$	$E_6 = IS + (QS/4)(0.234 + 0.940\eta^2 - 1.570\eta^4)$	10/84
$1/2 \rightarrow 3/2$	$E_7 = IS + (QS/4)(-0.743 - 1.086\eta^2 + 1.549\eta^4)$	3/84
$1/2 \rightarrow 1/2$	$E_8 = IS + (QS/4)(-0.143 - 0.430\eta^2 + 1.168\eta^4)$	18/84

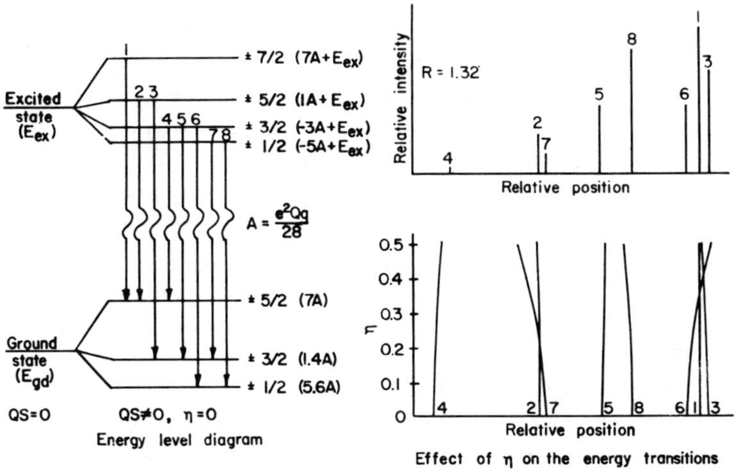

Fig. 2. Sb^{121} quadrupole interaction.

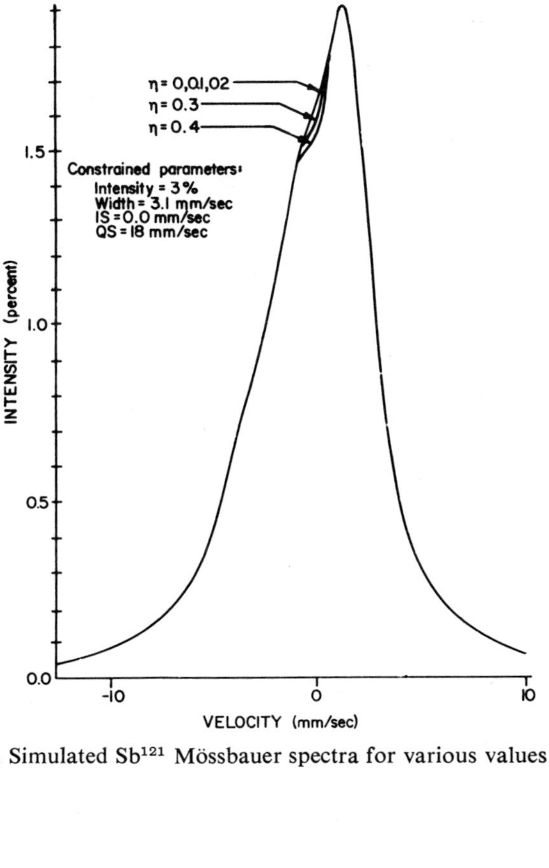

Fig. 3. Simulated Sb^{121} Mössbauer spectra for various values of η.

the effect of η on the transitions. As seen in the figure, η does affect the positions of the lines to some extent.

The transition energies (Table IV) were used to simulate Sb^{121} Mössbauer spectra for different values of η to determine if there would be noticeable change in the shape of the spectrum (Fig. 3). Results indicate an effect on the shape of the spectrum for $\eta > 0.2$, but very little effect for $\eta < 0.2$.

Computer results for the Mössbauer spectrum of orthorhombic Sb_2O_3 were obtained with $\eta = 0$ and with η as a parameter in the QS interaction (Fig. 4). A better fit was achieved when η was allowed to vary. The value obtained for η was 0.51 ± 0.14, within experimental error of that reported previously (Table II).

The Mössbauer spectrum of α-Sb_2O_4 offers direct confirmation of the conclusions of Skapski and Rogers [7] in their X-ray studies that this compound is actually a mixed oxidation state [Sb(III), Sb(V)], rather than Sb(IV). The spectrum in Fig. 5 shows two completely resolved peaks, one observed at a velocity associated with the +3

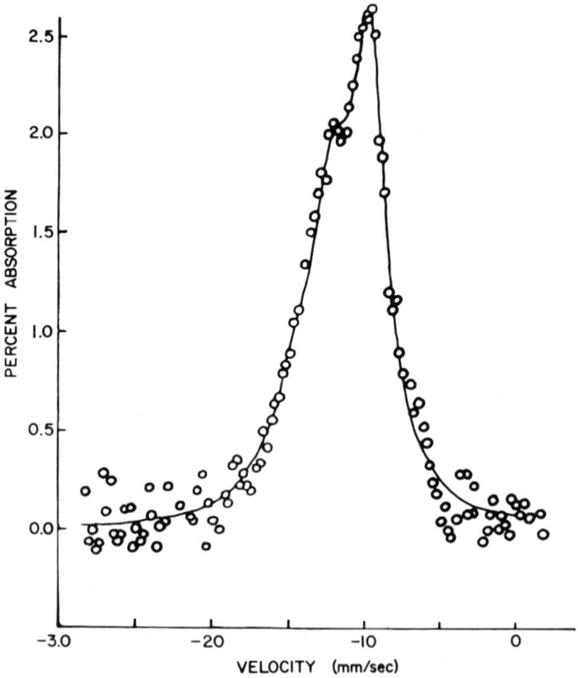

Fig. 4. Mössbauer spectrum of orthorhombic Sb_2O_3.

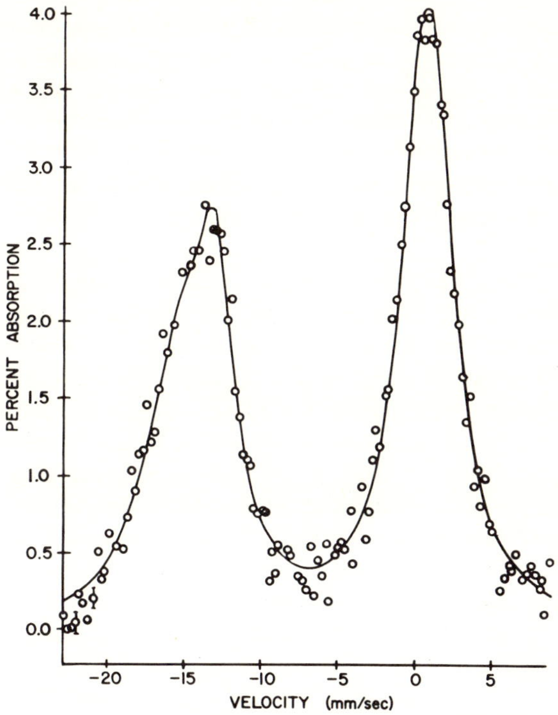

Fig. 5. Mössbauer spectrum of α-Sb_2O_4.

state, and the other at a velocity for the +5 state. It should also be noted that the Sb(III) peak has an *IS* (−14.36 ± 0.07 mm/sec) differing from the experimental value for cubic Sb_2O_3 (−11.32 ± 0.12 mm/sec), but the Sb(V) *IS* (0.61 ± 0.03 mm/sec) lies close to the experimental value for Sb_2O_5 (1.06 ± 0.02 mm/sec). The difference in *IS* for the Sb(III)'s implies that the bonding around the Sb^{3+} is different in α-Sb_2O_4 from that in cubic Sb_2O_3. This was expected on the basis of X-ray structural analysis of these two compounds [5, 7]. Each antimony in cubic Sb_2O_3 is surrounded by three oxygens in a pyramidal configuration [4]. α-Sb_2O_4 is isomorphous with $SbNbO_4$, in which the configuration about Sb(III) has two oxygens with an angle of ~90° and two oxygens farther away with an angle of ~150° [7]. The *IS* corresponding to the Sb(III) peak in the α-Sb_2O_4 spectrum corresponds to an increase in *s* electron density at the nucleus and a decrease in *s* character of the bonds when compared with the cubic

Sb_2O_3. This is not surprising, since the two close neighbors are at an angle of $\sim 90°$, implying very little or no *s* character in these bonds. Since the next two neighbors are positioned farther away, these bonds are weaker and probably have some *d* character as well. They are therefore not as efficient in withdrawing *s* electrons from the antimony. The agreement in *IS* suggests that the electron density, and possibly the bonding around Sb(V), might be similar in α-Sb_2O_4 and Sb_2O_5.

An asymmetrical peak was observed for both Sb_2O_5 and the Sb(V) in α-Sb_2O_4, suggesting the existence of either *QS* or more than one antimony site. The *IS*, as was pointed out above, is similar to that of the Sb(V) peak of α-Sb_2O_4. The symmetry around the Sb(V) of α-Sb_2O_4 is presumed to be a distorted octahedron [7]. The site geometry about Sb in Sb_2O_5 is not known. The difficulty of interpreting slight asymmetry in the Mössbauer peaks as being due to *QS* or multiple sites precludes a decision on bonding from these results alone.

The value for *QS* for amorphous Sb_2O_3 is within the experimental error for the value of cubic Sb_2O_3, and η was not required for fitting, which implies that the local environment about Sb is axially symmetrical in the amorphous material and similar to the Sb_4O_6 molecule in cubic Sb_2O_3.

Sulfides of Antimony

The amount (in percent) of antimony in each of the samples was determined by permanganate titration [8]. A Soxhlet extraction with CS_2 was used to remove free crystalline sulfur. The data are summarized in Table V. It is noted that of the commercial samples of Sb_2S_5, only sample III had the correct stoichiometry, and even that sample lost sulfur upon extraction.

The Mössbauer spectra show three interesting results. First, the isomer shifts of the "antimony pentasulfides" (I, II, and III) are within the velocity range of other Sb^{3+} compounds that have been observed in this (Table I) and previous investigations [1–3]. Therefore, the antimony in these compounds has an oxidation state of +3 and not +5. Secondly, the *IS* for amorphous Sb_2S_3 and the Sb_2S_5 compounds is greater than that of stibnite. This increase in *IS* corresponds to a decrease of *s* electron density at the nucleus and suggests some sort of change in the chemical bonding.

The decrease in *s* electron density at the nucleus of the antimony for the amorphous compounds agrees with the closer Sb–S bond

Table V. Analytical Data for Antimony Sulfides

Sulfide	Per cent Sb*	Formula (Sb_2S_n) for compound	Per cent weight loss from extraction	Formula (Sb_2S_n) after extraction	Linewidth from Mössbauer spectrum, mm/sec
Stinite	71.9	$Sb_2S_{2.97}$	—	—	4.31 ± 0.17
Amorphous antimony trisulfide	68.1	$Sb_2S_{3.55}$	0.0	$Sb_2S_{3.55}$	5.22 ± 0.18
Antimony pentasulfide(I)†	63.2	$Sb_2S_{4.42}$	7.2	$Sb_2S_{4.00}$	5.75 ± 0.37
Antimony pentasulfide(II)‡	23.8	$Sb_2Sb_{24.3}$	24.9	$Sb_2S_{16.1}$§	5.73 ± 0.26
Antimony pentasulfide(III)¶	60.1	$Sb_2S_{5.04}$	8.2	$Sb_2S_{4.00}$	5.27 ± 0.29

*Determined by permanganate titration [8].
†Sample from Alfa Inorganics, Inc., Cat. No. Sb–117.
‡Sample from E. H. Sargent and Co.
§Extraction was extended over six days with weight still not constant.
¶Sample from Pfaltz and Bauer Co.

distances in the amorphous material compared to stibnite, as inferred from electron diffraction studies [9].

The third result to be noted is the broadened linewidth. X-ray analysis of stibnite has shown two Sb sites [10], and there is also *QS*, as determined by NQR [11], although only one set of lines was measured. Both of these factors would increase the observed linewidth of the Mössbauer spectra. The observed linewidth was 4.31 ± 0.17 mm/sec, which is almost twice that of the natural linewidth and ~1.4 times larger than the oxides. The linewidth broadens from stibnite to amorphous Sb_2S_3 to Sb_2S_5 (Table V). This can be interpreted as an increase in the number of slightly different sites.

Halides of Antimony

Mössbauer spectra were obtained for all the simple antimony halides. However, in most cases the spectra were difficult to obtain due to the reactivity of the samples and the low intensity of the absorption.

The pentafluoride spectrum consists of a low-intensity peak. Although liquid SbF_5 is reported to have octahedral coordination [16], the structure of the solid SbF_5 is unknown. The *IS* of solid SbF_5

[2.23 ± 0.12 mm/sec (this work)] is between that of $NaSbF_6$ (2.0 ± 0.2 [3]) and that of $KSbF_6$ (3.7 ± 0.2 mm/sec [2]). No definite conclusions on structure can be made from the Mössbauer spectrum. The two probable configurations are octahedral coordination or trigonal bipyramidal coordination of the fluorides about the antimony.

The structure of the low-temperature phase of $SbCl_5$ is unknown. Two structures have been suggested: a dimeric arrangement in which each antimony is surrounded by six chlorines [13, 14], and an ionic structure of $SbCl_4^+ + SbCl_6^-$ [14]. The latter has at least two distinct sites for antimony. The *IS* of $SbCl_5$ [−3.12 ± 0.03 mm/sec (this work)] and $HSbCl_6 \cdot nH_2O$ (−3.0 ± 0.2 mm/sec [3]) are similar.

The Mössbauer results for both antimony(V) halides are consistent with octahedral symmetry about Sb similar to that in $SbCl_6^-$ and SbF_6^- (the latter containing essentially ionic Sb^{5+}). On this basis alone, however, one cannot rule out other coordination numbers, as it is not known how much effect on the value of *IS* a change in coordination from, for example, $SbCl_6^-$ to $SbCl_4^+$ would have. In fact, an analysis of the $SbCl_5$ spectrum into two Lorentzian peaks gave equally good fit to the data. The peaks were of approximately equal intensity and differed in *IS* by about 1 mm/sec. This could be considered in agreement with a two-site structure, if the difference in coordination has that small an effect on *IS*.

A linear relationship between *IS* and the electronegativity difference ΔX [16] was observed for the trihalides of antimony and cubic Sb_2O_3. Such a relationship has also been observed for the tetrahalides of tin. Numerous interpretations of the tin case can be found in the literature which usually relate the results qualitatively or quantitatively to the ionic character [17–19] or to the type of bonding [20–23].

The plot of *IS versus* ΔX is given in Fig. 6. The least-squares fit of the data gives the empirical equation $IS = -18.3 + 4.35\,\Delta X$, where the units for *IS* are mm/sec.

A linear plot (Fig. 7) is also obtained for *IS versus QS* when an adjusted *QS* value for SbI_3 is used. This value was obtained from NQR [24], noting the experimental *QS* value was smaller than would be expected. X-ray structural data on SbI_3 [25] show that the configuration about the antimony approaches octahedral, which will cause a decreased *QS*. On the plot [24] of *QS* (from NQR) *versus* ΔX for Sb(III) compounds ($SbCl_3$, $SbBr_3$, SbI_3, $SbI_3 \cdot 3S_8$) a projected value for SbI_3 can be obtained. This value of ~250 MHz is used for

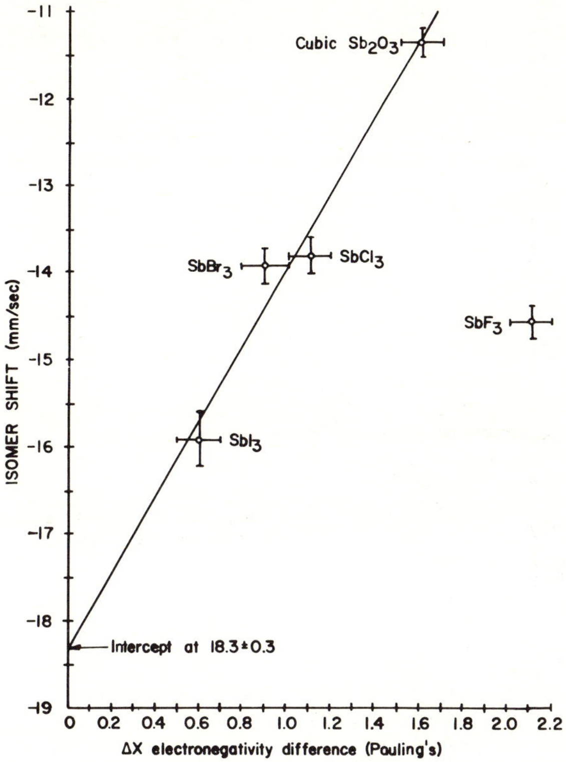

Fig. 6. Plot of isomer shift (*IS*) *vs.* electronegativity difference (ΔX) of some Sb(III) compounds.

the MES plot of *IS versus QS*. The empirical formula obtained from a least-squares fit is $IS = -19.2 + 0.0143QS$, where the units for *IS* are mm/sec and for *QS* are MHz.

The linearity of these two plots implies that the *s* character is constant in this series, which is expected due to the bond angles being approximately the same [4, 25–27]. The NQR results [24] have also been interpreted in terms of constant *s* character. The positive slope of the *IS versus* ΔX plot implies that the bonding about the antimony is becoming more ionic in going from $SbBr_3$ to Sb_2O_3. The SbI_3 falls on the line for the following reasons. It is the only compound in which the Sb–X distance is greater than the sum of the covalent radii. This would cause an increase in *s* electron density at the nucleus. However, the next-nearest-neighbors are close enough so that the Sb

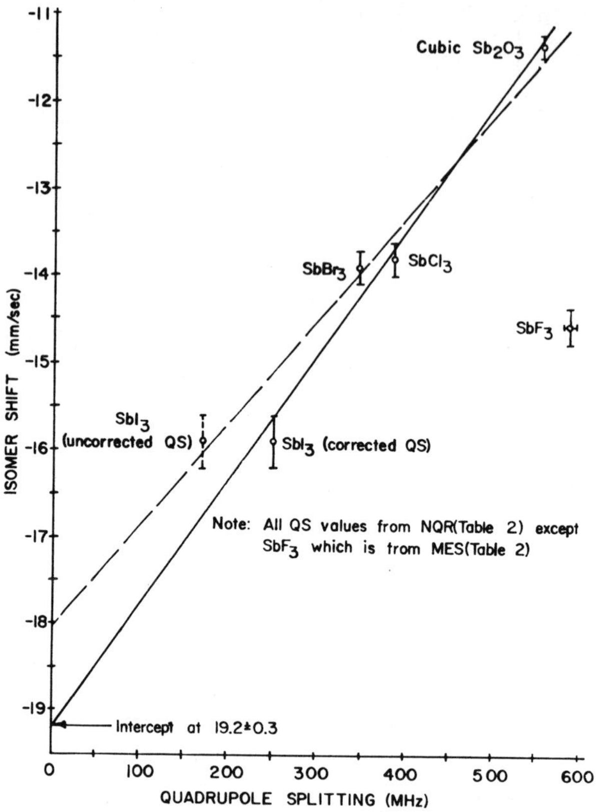

Fig. 7. Plot of isomer shift (*IS*) *vs.* quadrupole splitting constants (e^2qQ) of some Sb(III) compounds.

is in a displaced center of an octahedron. This additional number of neighbors tends to decrease the electron density. From the plot *IS versus* ΔX it seems that these two effects have canceled each other.

The slope of the line in Fig. 6 is ~4 mm/sec. This is compared with the slope for the Sb(V) compounds, which is ~7 mm/sec (Fig. 8). The slope is related to the total *s* character of all the bonds. The total *s* character is greater for the Sb(V) compounds, in which d^2sp^3 bonding presumably occurs. The bonding in the Sb(III) compounds is somewhat between sp^3 and p^3 bonds.

The *s* character for the plotted Sb(III) compounds (Fig. 6) is determined by obtaining the value of *IS* extrapolated to $\Delta X = 0$, by assuming that the *s* character for the bonds in InSb is 25% (InSb

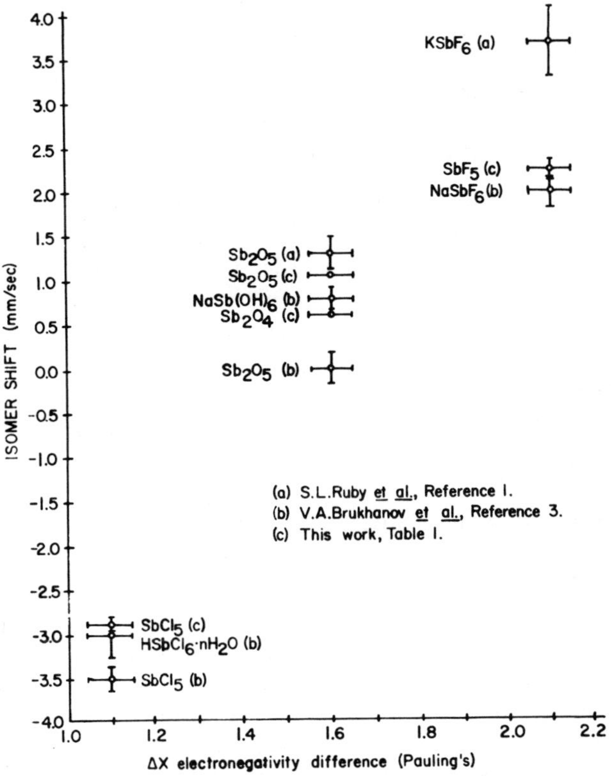

Fig. 8. Plot of isomer shift (*IS*) *vs*. electronegativity difference (ΔX) of some Sb(V) compounds.

has tetrahedral bonds) and by using zero *s*-bond character for pure ionic Sb(III), Ruby *et al.* [2] have calculated the *IS* value for ionic Sb(III). Using the extrapolated *IS*, the *IS* for InSb, the *IS* for pure ionic Sb(III), and assuming a linear relationship between *IS* and *s* character, a value of 14 $\pm$ 2% *s*-bond character for the trivalent series of Sb(III) compounds (SbI_3, $SbBr_3$, $SbCl_3$, and cubic Sb_2O_3) is determined. This can be compared with the value of 8% obtained from NQR [24.]

The *IS*–ΔX point for SbF_3 in the plot of Fig. 6 does not lie near the line, the *IS* being more negative. Therefore, the bonding in SbF_3 must be different from the bonding in these compounds. The implication is that the bonding in SbF_3 has less *s* character and more *p* character than the other trihalides of antimony. The bond angle between the two closest neighbors is less than 90°, indicating *s* character of almost zero for these bonds.

The QS in these trihalides is large because of the pyramidal structure about the antimony, which creates a large electric-field gradient at the antimony nucleus.

SUMMARY AND CONCLUSIONS

Using cubic Sb_2O_3 it was found that $R = 1.32 \pm 0.02$, which compares to the value of 1.38 ± 0.02 given by Ruby *et al.* [1]. An accurate value of R is needed in order to obtain accurate values for both QS and η, and all of these parameters, along with linewidth and intensity, are required to fit the observed spectrum. Additional work is definitely needed to evaluate R more exactly. In most cases η will have no observable effect on the spectrum, but when η reaches values >0.3, a noticeable change in the spectral shape occurs. Values of η obtained by NQR are several orders of magnitude more precise than can be obtained by the Mössbauer technique.

Also affecting the shape is QS. The QS is needed in order to determine accurate values of IS. Again, as with η, the values for QS obtained from (MES) do not have nearly the precision that NQR results have.

Interpretation of the Sb^{121} Mössbauer spectra in structural terms at this stage of development appears uncertain without the structural information available, preferably from X-ray analysis. However, oxidation states can be determined easily, and in this respect Sb^{121} MES can contribute to understanding the nature of a number of compounds. The IS values for the Sb_2S_5 samples studied show the antimony to be in the +3 oxidation state, unlike Sb_2O_5, which is +5. Also, IS values have confirmed α-Sb_2O_4 to be a mixed oxidation state [Sb(III), Sb(V)].

A linear relationship between electronegativity difference ΔX and IS was observed for a series of Sb(III) compounds (SbI_3, $SbBr_3$, $SbCl_3$, and cubic Sb_2O_3), as well as for all the reported Sb(V) compounds. A plot of IS *versus* QS for the Sb(III) compounds is also linear. The positive slope of these lines suggests the bonding has some s character. The bonding in the trihalides, except SbF_3, has about 14% s character. The s character in SbF_3 is less than that of the other trihalides.

ACKNOWLEDGMENTS

The authors are indebted to G. G. Long for his assistance in this work and for many helpful discussions on antimony chemistry. They also appreciate that S. L. Ruby of Argonne National Laboratory provided several samples of antimony sulfides, and as well communicated his unpublished isomer-shift data on these. The authors would like to thank R. J. Tullbane for the preparation of the Sb_2O_3 modifications and H. Z. Dokuzoguz for X-ray diffraction analysis of some of the compounds. The source was irradiated through the cooperation of the DuPont Savannah River Plant and the US Atomic Energy Commission. This work was partially supported by the National Science Foundation, Grant GP-8087. J. G. Stevens gratefully acknowledges the support of a NASA traineeship.

REFERENCES

1. S. L. Ruby, G. M. Kalvius, R. E. Snyder, and G. B. Beard, *Phys. Rev.* **148**:176 (1966).
2. S. L. Ruby, G. M. Kalvius, G. B. Beard, and R. E. Snyder, *Phys. Rev.* **159**:239 (1967).
3. V. A. Brukhanov, B. Z. Iofa, V. Kotkhekar, S. I. Semenov, and V. S. Shpinel, *Soviet Phys.—JETP* **26**:912 (1968).
4. T. N. Polynova and M. A. Porai-Koshits, *J. Struct. Chem.* (*USSR*) **7**:147 (1966).
5. R. M. Bozorth, *J. Am. Chem. Soc.* **45**:1621 (1923).
6. R. G. Barnes and P. J. Bray, *J. Chem. Phys.* **23**:1177 (1955).
7. A. C. Skapski and D. Rogers, *Chem. Commun.* **1965**:611.
8. W. C. Pierce, E. L. Haenisch, and D. T. Sawyer, *Quantitative Analysis* (John Wiley and Sons, New York, 1960), 4th ed., p. 283.
9. A. M. Reshetnikov, *Soviet Phys.—Cryst.* **4**:883 (1959).
10. S. Scavnicar, *Z. Krist.* **114**:85 (1960).
11. T. Wang, *Phys. Rev.* **99**:566 (1955).
12. C. J. Hoffman, B. E. Holder, and W. J. Jolly, *J. Phys. Chem.* **62**:364 (1958).
13. R. F. Schneider and J. V. Di Lorenzo, *J. Chem. Phys.* **47**:2343 (1967).
14. K. C. Olie, C. Smitskamp, and H. Gerding, *Inorg. Nucl. Chem. Letters* **4**:129 (1968).
15. D. J. Durrant and B. Durrant, *Introduction to Advanced Inorganic Chemistry*, (John Wiley and Sons, New York 1962), p. 767.
16. L. Pauling, *The Nature of the Chemical Bond* (Cornell University Press, Ithaca, New York 1960), 3rd ed., p. 93.
17. A. J. F. Boyle, D. St. D. Bunburny, and C. Edwards, *Proc. Phys. Soc.* (*London*) **79**:416 (1962).
18. V. S. Shpinel, V. A. Bryukhonov, and N. N. Delyagin, *Soviet Phys.—JETP* **14**:1256 (1961).
19. M. Cordey-Hayes, *J. Inorg. Nucl. Chem.* **26**:915 (1964).
20. J. K. Lees and P. A. Flinn, *Phys. Letters* **19**:186 (1965).
21. N. N. Greenwood, P. G. Perkins, and D. H. Wall, in: *The Mössbauer Effect*, (Symposium of the Faraday Society, No. 1, 1967) (Plenum Press, New York, 1968), p. 51.
22. V. I. Gol'Danskii, E. F. Makarov, and R. A. Stukan, *J. Chem. Phys.* **47**:4048 (1967).
23. J. K. Lees and P. A. Flinn, *J. Chem. Phys.* **48**:882 (1968).
24. S. Ogawa, *J. Phys. Soc. Japan* **13**:618 (1958).
25. J. Trotter and T. Zobel, *Z. Krist.* **123**:67 (1966).
26. I. Lindqvist and A. Niggli, *J. Inorg. Nucl. Chem.* **2**:345 (1956).
27. D. W. Cushen and R. Hulme, *J. Chem. Soc.* **1964**:4162.
28. I. A. Safin, *J. Struct. Chem.* (USSR). **4**:242 (1963).
29. T. P. Das and E. L. Hahn, *Solid State Phys.* **1959** (Suppl. 1):196.
30. R. Bersohn, *J. Chem. Phys.* **20**:1505 (1952).

Mössbauer Spectroscopy of Organometallic Compounds in Noncrystalline Matrices

S. Chandra and R. H. Herber

Wright Laboratory
Rutgers—The State University
New Brunswick, New Jersey

The frozen-glassy-matrix technique has been employed in a study of the lattice dynamics of a number of organo-iron and organotin compounds. This technique has also been exploited in determining possible conformational changes on dissolution of the crystalline solids in solvents which set to a glassy matrix at cryogenic temperatures. The choice of solvent primarily depends on (1) *a reasonably high lattice temperature compared to* kT (2) *nonreactivity with the solute under study, and* (3) *the property of forming a glassy* (*noncrystalline*) *structure on cooling. Mössbauer studies on* $C_8H_8Fe(CO)_3$, $[C_5H_5Fe(CO)_2]_2SnCl_2$, *and* $C_5H_5Fe(CO)_2Sn(C_6H_5)_3$ *as neat solids and as solutes in polymethylmethacrylate in the temperature range* $298 \geq T \geq 78°K$ *are reported. Far-infrared studies previously reported for these compounds show the presence of low-lying optical modes at about* $100\ cm^{-1}$. *By studying both* Fe^{57} *and* Sn^{119} *spectra the contribution of the optical modes toward the Mössbauer fraction can be determined from the nonconstancy of the area ratio of two resonance spectra in different matrices.*

INTRODUCTION

Mössbauer spectroscopy is generally used in studying the properties of pure solids or alloys. Relatively little has been reported concerning the study of frozen solutions by this technique, presumably due to some of the difficulties involved; namely, small resonance effects, reactivity of the solvents, and the low concentration of the resonance atoms. In the past few years Mössbauer spectroscopy has been extensively used in elucidating the structure and bonding of organometallic compounds of iron and tin [1]. Recently [2] this technique has been exploited in a unique manner in the study of the possible conformational changes* of organometallic compounds when these are subjected to changes in environment as in solution processes. The need for such studies lies in the fact that the information concerning the structure of chemical species obtained from X-ray diffraction data refer to that found in single crystals, while the structural information derived from other spectroscopic techniques, particularly infrared, nuclear magnetic resonance, and Raman methods, is generally obtained from solution samples. Since Mössbauer measurements can be carried out both on neat solids and frozen solutions, this method is able to bridge the gap between diffraction and spectroscopic measurements, and possibly resolve some of the apparent differences in the structural information so derived. In addition to answering the question of retention of conformation in the liquid state, Mössbauer spectroscopy in frozen solvents can also be used in the studies of solvent–solute effects or polymer formation. A typical example of the latter may be found in the work of Stöckler *et al.* [3], who have found a relatively slow decrease in the recoilless fraction in the temperature range 80–300°K in bis(toluene 3,4-dithiol tin) and methyl tin formate, as compared to other organotin compounds such as $(C_6H_5)_4Sn$ and $[(C_6H_5)_3Sn]_2$. This behavior could be related to the polymeric structure of the former molecules, and it was suggested that this polymeric type of molecular structure contributes to a large recoilless fraction.

The studies on the temperature dependence of the recoil-free fraction described above were made exclusively on solids. The present paper deals primarily with the technique of using frozen solutions. This technique has been utilized as a "micro" tool in studying the lattice dynamics of materials, especially with a view to elucidating the mixing of the local (optical) and acoustic vibrational modes. This mixing will possibly be pronounced in molecular crystals where both intra- and intermolecular interactions are present.

There are three kinds of solvents which can be used for this "frozen" solution technique:

1. Noncrystalline, glass-forming matrices: These solvents are glass-formers at low temperature; i.e., they accommodate the structure

*Conformational changes are those which involve changes in bond angles and/or rotation about one or more symmetry axes without bond breakage. Two or more distinguishable structures of this type are usually referred to as "conformers" or "conformational isomers" (see also Ref. [18]).

of the solute rather than imposing a crystalline environment on the latter. Some examples are EPA (16 % V/V ethanol, 42 % V/V ipropanol, and diethyl ether), methyl pentane, methyl hydrofuran, and polymethylmethacrylate. The solvent chosen for Mössbauer studies must meet several other requirements, namely, chemical inertness and good solvating power for the solutes in question, and should have a reasonably high lattice temperature (so that the resonance effect will be as large as possible at the temperatures of measurement). Polymethylmethacrylate is especially suitable in the latter aspect.

2. Frozen inert-gas matrices like argon: These are useful for studying isolated free molecules in which chemical interaction with the matrix can be ignored. Data for such systems should be relatively easy to interpret. A typical example of such a study is the recently reported [4] determination of the Mössbauer spectrum of I_2 in various matrices, including solid argon.

3. Crystalline matrices: Mössbauer spectroscopy of frozen aqueous solutions and nonaqueous solutions such as benzene and acetonitrile is useful in the investigation of (1) ions "trapped" in ice, (2) complexes in solution, and (3) phase transitions. Nozik and Kaplan [5] have carried out experiments on frozen aqueous solutions of ferrous chloride and ferrous sulphate. They have observed several phase changes in the ice–salt systems and are able to estimate the symmetry properties of the trapped ion and the changes in symmetry which result from transformations in the ice structure.

EXPERIMENTAL

As part of a continuing study of structure and bonding of organometallic compounds, we have carried out measurements on a number of such materials both as neat solids as well as in frozen solutions. Frozen-solution samples were prepared by dissolving the sample compound in an appropriate amount of solvent (EPA, MTF, methylpentane) and then transferring this material to a liquid sample cell by means of a hypodermic-syringe injection system. The sample cell, which consisted of a solution volume of about 40 mil thickness bounded by two 5-mil-thick Mylar windows, was immersed immediately after filling into liquid nitrogen and maintained at this temperature until mounted in the experimental Dewar for measurements. Dilute solutions in polymethylmethacrylate were prepared by mixing a benzene solution of a known amount of the polymer and of the sample in question, transferring this to cylindrical cups of Teflon, and allowing the excess solvent to evaporate at room temperature. The resultant disks were stripped off, trimmed if necessary to remove excess material, and then mounted in stacks in the usual sample position of the liquid-nitrogen Dewar system. A Cu–Au(Co) thermocouple was used to measure the temperature at the sample holder. An automatic temperature regulator, in conjunction with a carbon resistor as temperature sensor, gave a temperature regulation of $\pm 2°K$ in the range $80 < T < 220°K$. The Mössbauer spectrometer using a parabolic drive (linear velocity) system has been described previously. A source of Co^{57} diffused into palladium was used for the Fe^{57} resonance (14.4-keV gamma rays) and pertinent isomer shifts are reported with respect to the centroid of a standard sodium nitroprusside absorber spectrum at room temperature. The 23.8 keV resonance of Sn^{119} was observed using a barium stannate source, and the pertinent isomer shifts are reported with respect to the centroid of a stannic oxide or barium stannate spectrum at room temperature. The K X-ray was filtered out by using a 0.001-in.-thick palladium foil.

In certain situations, especially for glassy-matrix absorbers, where the spectra show small resonance effects (1–2 %), the counting was continued until $\sim 10^6$ counts per channel had been accumulated, so that the statistical counting error was reduced to $<0.1\%$ for a signal-to-noise ratio of >10.

In the present studies we were not concerned with the absolute values of the recoilless fraction f. The variation of f with temperature was determined by measuring the temperature dependence of the area A under the resonance curve ($A = \frac{1}{2}\pi\,\epsilon\Gamma$, where ϵ is the percentage absorption at resonance maximum and Γ is the linewidth at half maximum). This correspondence between $d(\ln f)/dt$ and $d(\ln A)/dt$ is justified in the case of "thin" absorbers, where saturation effects are presumably absent.

Optical-Mode Dependence of Mössbauer *f* Values

Assuming a Debye model for the frequency spectrum, in a Mössbauer experiment the recoilfree fraction is given by [6, 7]

$$f = \exp\left\{-\frac{3}{2}\frac{E_R}{k\theta}\left[1 + 4\left(\frac{T}{\theta}\right)^2 \int_0^{\theta/T} \frac{x}{e^x - 1}\,dx\right]\right\} \qquad (1)$$

where E_R is the recoil energy of the free nucleus, θ is the Debye temperature, and $x = h\omega/2\pi kT$.

At higher temperatures, $T \geq \theta/2$, the integral in Eq. (1) approaches θ/T, so that

$$f = \exp -\left\{\frac{3}{2}\frac{E_R}{k\theta} + \frac{6E_R}{k\theta^2}T\right\} \quad (2)$$

$$d(\ln f)/dT = -6E_{R}/k\theta^2 \quad (3)$$

or

$$\theta = -\left[\frac{3E_\gamma^2}{2MC}\left(\frac{1}{d(\ln f)/dT}\right)\right]^{\frac{1}{2}} \quad (4)$$

Thus, θ for a given lattice can be determined from temperature-dependence measurements on f.

For a real system of polyatomic molecules in which the Mössbauer atom does not fulfill the requirements of the Debye model, the temperature so calculated should be identified as a "Mössbauer temperature," to be referred to as θ_M in the following discussion. The vibrational frequency ($\nu_m = k\theta_M/h$) is related to the Mössbauer–Lamb factor f not only through the value of θ_M, but also in terms of the mean-square amplitude of vibration $\langle X^2 \rangle$, since

$$f = \exp(-4\pi^2/\lambda^2)\langle X^2 \rangle \quad (5)$$

where λ is the equivalent wavelength of the Mössbauer gamma ray. If the vibrational amplitude of the Mössbauer atom rises only from vibrational modes of the whole crystal lattice assembly, the excitation is considered to be acoustical. On the other hand, we may speculate on what kind of local (optical) vibrational modes are needed to account for the temperature dependence of the recoilless fraction. From the Einstein relation $h\nu_M = k\theta_M$ it is seen that if the lowest-lying optical mode which involves motion of the Mössbauer atom occurs at 100 cm^{-1}, θ_M is approximately 144°K.

Let us now consider a molecular solid which contains two distinct Mössbauer atoms A and B, which are linked to each other through normal chemical bonds. The ratio of the recoilless fractions for the two atoms is given from Eq. (2) in the high-temperature limit by

$$\frac{\ln f_A}{\ln f_B} \approx \frac{3E_{R,A}/2k\theta_A + 6E_{R,A}T/k\theta_A^2}{3E_{R,B}/2k\theta_B + 6E_{R,B}T/k\theta_B^2} \quad (6)$$

which for $T/\theta \geq 1$ reduces to

$$\frac{\ln f_A}{\ln f_B} \approx \frac{E_{R,A}/\theta_A^2}{E_{R,B}/\theta_B^2} \quad (7)$$

If only acoustical modes contribute to the effective lattice temperature under the conditions of the measurement, then the lattice temperature will be same at A sites and at B sites, i.e., $\theta_A = \theta_B$, so that Eq. (7) becomes

$$\ln f_A / \ln f_B = E_{R,A} / E_{R,B} \tag{8}$$

i.e., the ratio of the recoil-free fractions is independent of the lattice temperature and is given by the ratio of the free-atom recoil energies. On the other hand, if local (optical) vibrational modes make a significant contribution to the mean-square amplitude of vibration, then $\theta_A \neq \theta_B$, and hence $\ln f_A/\ln f_B$ will depend on the nature of the matrix in which the Mössbauer atoms are embedded.

RESULTS AND DISCUSSION

Detailed Mössbauer studies have been carried out on cyclooctatetraene iron tricarbonyl [$COTFe(CO)_3$]. This compound, which was first reported in 1959 [8–10], belongs to the family of those "fluxional" organometallic molecules which have the following dynamical characteristics: (1) they have two or more chemically equivalent nuclear configurations with respect to the bonding of the metal-containing moiety; (2) under normally accessible experimental conditions these configurations pass from one to another through intermediate states which are accessible *via* low-energy pathways; (3) this interconversion, known as site exchange, proceeds rapidly enough to have detectable consequences in experimental measurements, particularly (though not exclusively) NMR spectra. The room-temperature NMR spectrum [9–11] of the $C_8H_8Fe(CO)_3$ molecule consists of a single sharp proton resonance, although the X-ray data [12] for the neat solid show that the iron atom is bonded to only four of the eight carbon atoms of the COT ring. Moreover, the COT protons in the iron compound are moved up field by about 70 ppm relative to the free COT ring, showing significant bonding of the metal to the ring moiety. The observed NMR singlet was interpreted as due to the fast (on the NMR time scale) averaging of the eight protons. The low-temperature NMR spectra [13–15] in appropriate solvents show a splitting of the proton resonance into two asymmetrical peaks symmetrically displaced from the room-temperature resonance. On the basis of the low-temperature NMR results, three possible structures were postulated as describing the conformation of $C_8H_8Fe(CO)_3$ in solution: (1) a biplanar COT ring acting as a

1,3-diene [13], (2) a tub-shaped COT ring acting as a 1,3-diene [14], and (3) a tub-shaped COT ring acting as a 1,5-diene [15].

The apparent disagreement concerning the structure of the low-temperature solution conformation of COT $Fe(CO)_3$ and its relationship to the structure of the neat solid has been resolved by carrying out Mössbauer measurements on the neat solid and on frozen solutions of COT $Fe(CO)_3$ in glassy-matrix-formers such as EPA, MTF, and PMM (Fig. 1), as well as *n*-octane and nitrobenzene [2]. The Mössbauer parameters (Table 1) were found to be essentially the same for the solid and in frozen glassy matrices. For this molecule the quadrupole splitting (which is most sensitive to variations in the structures which are being considered) is markedly different from that observed for $C_8H_{12}Fe(CO)_3$, which is known to have a 1,5-diene structure, and very much smaller than the *QS* of norbornadienyl iron tricarbonyl $C_7H_8Fe(CO)_3$, in which the metal atom is bonded to a nonalternate cyclic diolefin in which the two double bonds are in a 1,4 position relative to each other. By comparing the *QS* with those observed for related compounds as well as a consideration of related infrared and NMR data it was concluded [16] that COT $Fe(CO)_3$ in liquids has a 1,3-diene structure similar to that originally determined for the solid.

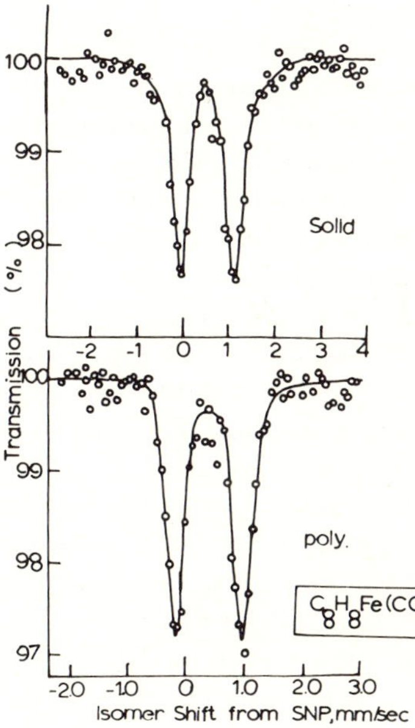

Fig. 1. Mössbauer spectra of $C_8H_8Fe(CO)_3$ as neat solid and in PMM at 80°K. The isomer shift is with respect to SNP at 294°K.

TABLE I

Compound	Temp., °K	Matrix	Resonance	IS,* mm/sec	QS, mm/sec
$[C_8H_8Fe(CO)_3]$	80	Solid	Fe	+0.314	1.240
	80	PMM	Fe	+0.375	1.260
	80	EPA	Fe	+0.340	1.16
	80	MTF	Fe	+0.350	1.20
$[(C_5H_5)Fe(CO)_2]_2SnCl_2$	80	Solid	Fe	+0.338	1.682
	80	PMM	Fe	+0.351	1.743
	80	Solid	Sn	+1.959	2.248
	80	PMM	Sn	+1.62	2.92
				+1.65	1.62
				+1.96	2.25
				+1.31	2.29
$[(C_5H_3)Fe(CO)_2]Sn(C_6H_5)_3$	120	Solid	Fe	+0.316	1.925
	120	PMM	Fe	+0.341	1.870
	120	Solid	Sn	+1.485	0
	120	PMM	Sn	+1.50	0
$Fe_2(SnO_3)_3$	120	Solid	Fe	+0.768	0.736
	120	Solid	Sn	+0.055	0.545
$Sn_2Fe(CN)_6$	120	Solid	Fe	+0.164	0
	120	Solid	Sn	+3.162	1.370

*Isomer shifts for Sn^{119} and Fe^{57} are with respect to SnO_2 and $Na_2[Fe(CN)_5NO]\cdot 2H_2O$, respectively.

To clarify further the structure of the cyclooctatetraene iron tricarbonyl molecule both in the neat solid form as well as in solution, temperature dependence studies have been carried out. Such studies have been motivated by data obtained from far-infrared studies [17] on this molecule, which show the presence of a metal-ring bending mode at $\sim 100\ cm^{-1}$. Such low-lying optical modes—which are characteristic of a particular structural relationship between the C_8H_8 ring and the $Fe(CO)_3$ group—are expected to make an important contribution to the temperature dependence of the recoil-free fraction.

The temperature dependence of the area under the resonance doublet for $C_8H_8Fe(CO)_3$ in various matrices is summarized in the semilogarithmic plot given in Fig. 2. The characteristic Mössbauer temperature θ_M for the neat solid is 149°K and in polymethylmethacrylate it is 120°K. Furthermore, this temperature remains invariant at different solute–solvent ratios. These results must be examined in the light of the observation that there is an optical (local) mode of vibration present at $100\ cm^{-1}$ corresponding to a lattice temperature of 144°K. It seems unlikely that *only* this mode is populated to give

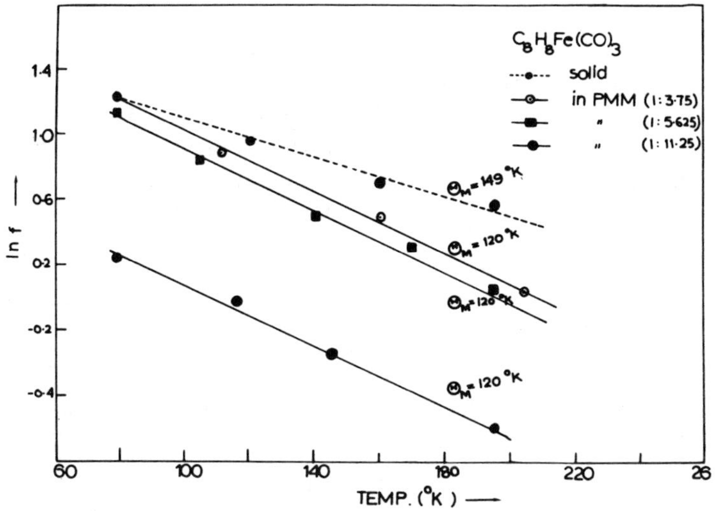

Fig. 2. Curves of $\ln f$ *vs.* T for $C_8H_8Fe(CO)_3$ as neat solid and in PMM (with solute–solvent ratio of 1 : 3.75, 1 : 5.625, and 1 : 11.25 by weight).

the Mössbauer temperature of 149°K; and the closeness of the two values (namely, θ_M and the lattice temperature corresponding to the optical mode of vibration present) may be fortuitous. Nevertheless, it does suggest a possible contribution of the optical modes of vibration of Mössbauer atoms in determining the temperature dependence of the recoilless fraction.

The observation that the Mössbauer temperature of COT $Fe(CO)_3$ in a frozen matrix of PMM is different from that observed in the neat solid suggests that the Mössbauer temperature so obtained is probing simultaneously the matrix and the existing low-lying optical modes. This has led us to a study of other organometallic molecules in order to relate θ_M with the presence or absence of low-lying local (optical) modes, keeping the nature of the supporting matrix fixed.*

DOUBLE MÖSSBAUER RESONANCE EXPERIMENTS

In order to understand further the role played by local optical modes in the lattice dynamics of these materials, we have studied a number of compounds which contain two distinct Mössbauer

*Of course, it would be interesting to study the behavior of the various matrices toward their recoilless fraction f by choosing a particular compound in which a low-lying optical mode is known to be present and dissolving this in various matrices and carrying out the appropriate temperature-dependence studies.

resonance nuclei. The compounds which were studied as neat solids as well as frozen solutions of PMM were $[\pi\text{-}C_5H_5Fe(CO)_2]_2SnCl_2$ and $[C_5H_5Fe(CO_2]Sn(C_6H_5)_3$ [18]. Two inorganic compounds, $Fe_2(SnO_3)_3$ and $Sn_2Fe(CN)_6$, were also studied as neat solids. Because of their limited solubility in nonreactive glass-forming solvents, it was not possible to study these materials in solution form. The results are shown in Tables I and II. In $Fe_2(SnO_3)_3$ the characteristic Mössbauer

Table II

Compound	Matrix	θ_M,°K	Mössbauer "Probe"
$C_8H_8Fe(CO)_3$	Neat solid	149	Fe^{57}
	PMM	120	Fe^{57}
$[C_5H_5Fe(CO)_2]_2SnCl_2$	Neat solid	180	Fe^{57}
	Neat solid	145	Sn^{119}
	PMM	132	Fe^{57}
	PMM	Rotational isomers	Sn^{119}
$[C_5H_5Fe(CO)_2]Sn(C_6H_5)_3$	Neat solid	133	Fe^{57}
	Neat solid	132	Sn^{119}
	PMM	—	Fe^{57}
	PMM	—	Sn^{119}
$Sn_2Fe(CN)_6$	Solid	177	Fe^{57}
	Solid	126	Sn^{119}
$Fe_2(SnO_3)_3$	Solid	206	Fe^{57}
	Solid	204	Sn^{119}

temperature,* as "probed" by the iron and tin atoms are, as expected, identical, which suggests the absence of local modes of vibration, and $\theta_M \sim 206$°K is taken to be a characteristic temperature of the whole lattice [Fig. 3]. In the case of stannous ferrocyanide it was expected that the same Mössbauer temperatures for Fe and Sn would be observed. Contrary to this expectation, however, the Mössbauer temperature is significantly different for the two sites, as shown in Fig. 4. Since far-infrared data on this compound are not available, it is not possible to carry out a detailed interpretation of these results, but further studies of this compound are currently underway.

The observation of identical Mössbauer temperatures for Fe and Sn in ferric stannate may be compared to the results of Bukshpan and Herber [19] for the temperature dependence of the recoilless fraction

*It should be noted that a linear dependence of $\ln f$ on the temperature does not *necessarily* imply a single characteristic frequency spectrum; however, it is useful to treat θ_M as a characteristic temperature of the material as probed by the particular Mössbauer atom.

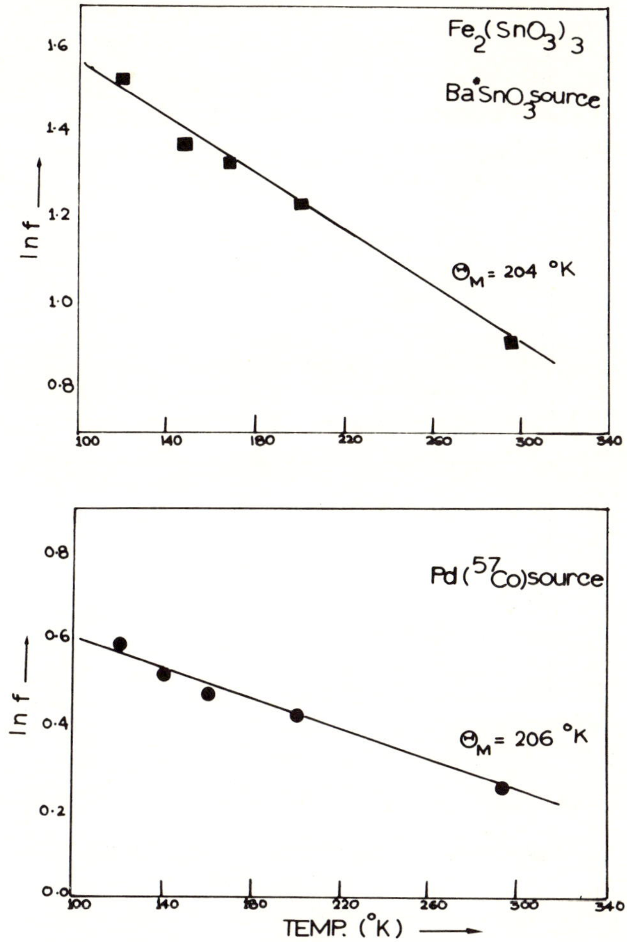

Fig. 3. Curves of ln *f* *vs.* *T* for $Fe_2(SnO_3)_3$ as neat solid: $BaSn^{119m}O_3$ and $Pd(Co^{57})$ sources.

derived from Sn^{119} and I^{129} experiments on stannic iodide, a molecular crystal with high local symmetry and well-characterized infrared and Raman-active optical modes. In this case two significantly different characteristic temperatures were found: 166°K for the Sn^{119} resonance and 85°K for I^{129} resonance. These results have been interpreted on the basis of the contribution to the vibrational modes of optically

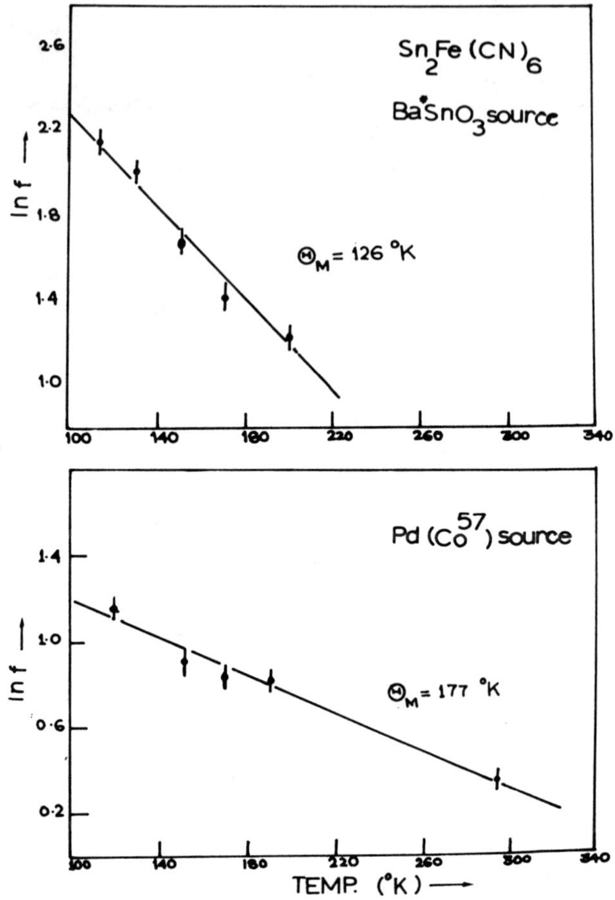

Fig. 4. Curves of ln *f vs. T* for $Sn_2Fe(CN)_6$ as neat solid: $BaSn^{119m}O_3$ and $Pd(Co^{57})$ sources.

active transitions which differ for the iodine and tin sites, in consonance with the calculated force constants for the molecule.*

In the case of the organometallic molecular compound $[C_5H_5Fe(CO)_2]_2SnCl_2$, where the presence of low-lying optical modes of vibration are expected, the characteristic Mössbauer temperatures

*Hazony [20] has analyzed these data on the temperature dependence of the Mössbauer fraction in terms of the inter- and intramolecular vibrations. The characteristic Einstein temperatures for the intermolecular translational and rotational vibrations were calculated as $\theta_t = 44$°K and $\theta_{rot} = 66$°K. The θ_t so obtained is very close to the observed optical (far-infrared) antitranslational vibration frequency (48°K) of the eight molecules in the cell.

in the neat solid (Fig. 5) are in fact found to be different for Fe and Sn atoms [θ_M(Fe) = 180°K; θ_M(Sn) = 145°K]. The Mössbauer temperature of the Sn^{119} resonance atoms of $[C_5H_5Fe(CO)_2]_2SnCl_2$ in PMM could not be uniquely determined, because of the existence of two conformationally related structures of this compound in solution (Fig. 6). Fe^{57} spectra of $[C_5H_5Fe(CO)_2]_2SnCl_2$ in PMM gave θ_M = 132°K (Fig. 7). In the light of the discussion given earlier for θ_M(Fe) = 120°K of $C_8H_8Fe(CO)_3$ in PMM, a possible explanation for a higher value of θ_M(Fe) = 132°K of $[C_5H_5Fe(CO_2)]_2SnCl_2$ in PMM may be that the *lowest* optical mode of vibration contributing to the recoilless fraction in the latter compound lies at a higher corresponding lattice temperature than in the former (notice that both compounds are in the same matrix).

In the case of $[C_5H_5Fe(CO)_2]Sn(C_6H_5)_3$, it was found that θ_M(Fe) = 133°K and θ_M(Sn) = 132°K (Fig 8).

CONFORMATIONAL STUDIES OF $C_5H_5Fe(CO)_2Sn(C_6H_5)_3$ AND $[C_5H_5Fe(CO)_2]_2SnCl_2$

In $C_5H_5Fe(CO)_2Sn(C_6H_5)_3$ there are no observable conformational changes on dissolution of the crystalline solids in solvents which

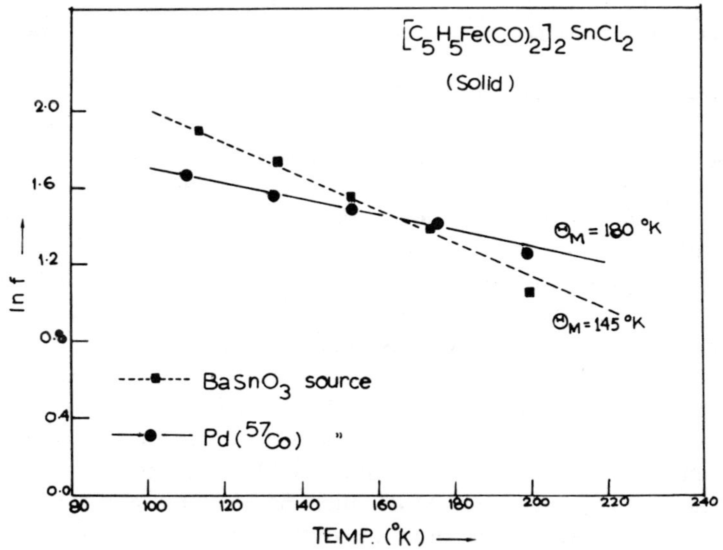

Fig. 5. Curves of ln *f* *vs.* *T* for $[C_5H_5Fe(CO)_2]_2SnCl_2$ as neat solid: $BaSn^{119m}O_3$ and $Pd(Co^{57})$ sources.

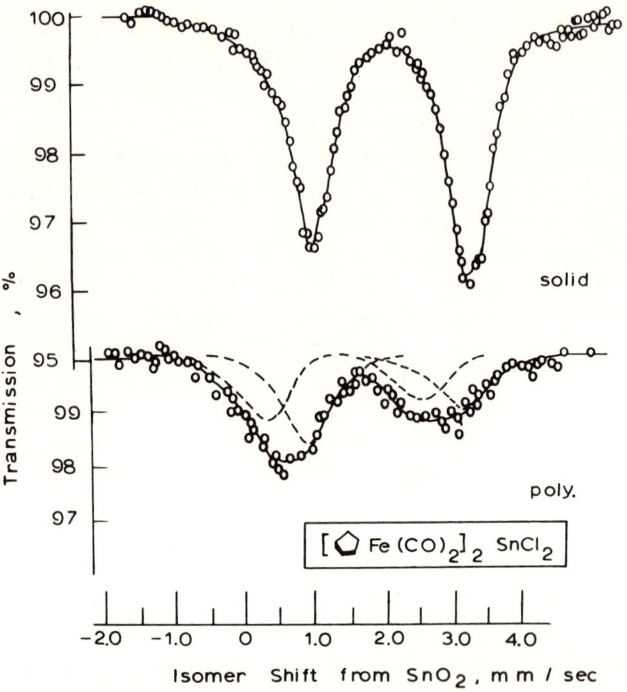

Fig. 6. Sn^{119} Mössbauer spectra of $[C_5H_5Fe(CO)_2]_2SnCl_2$ as neat solid and in PMM at 80°K. The isomer shift is with respect to SnO_2. $BaSn^{119m}O_3$ source.

set to a glassy matrix. The X-ray diffraction data on $[C_5H_5Fe(CO)_2]_2$ $SnCl_2$ reported by O'Connor and Corey [21] show that the Sn—Fe bond length of 2.492 Å is shorter than any previously reported tin–transition-metal bond length and the Sn—Cl bond length of 2.43 Å is longer than the metal–halogen bond lengths reported for $SnCl_4$, CH_3SnCl_3, etc. Moreover, the departure from tetrahedral values of the Cl—Sn—Cl bond angle (94.1°) and of the Fe—Sn—Fe bond angle (128.6°) is the largest yet observed for a four coordinate tin atom. It is possible that these anomalous values arise from the intermolecular forces which exist when the molecules are stacked into a solid crystalline matrix. On the other hand, if the bond angles and bond distances observed in the solid are consequences of the intramolecular bonding interaction, dissolution of the crystal would be expected to leave the structure—and hence the Mössbauer parameters—essentially unchanged. That this latter situation obtains can be readily inferred from the Fe^{57} data, for which both the *IS* and *QS* of the neat solid and

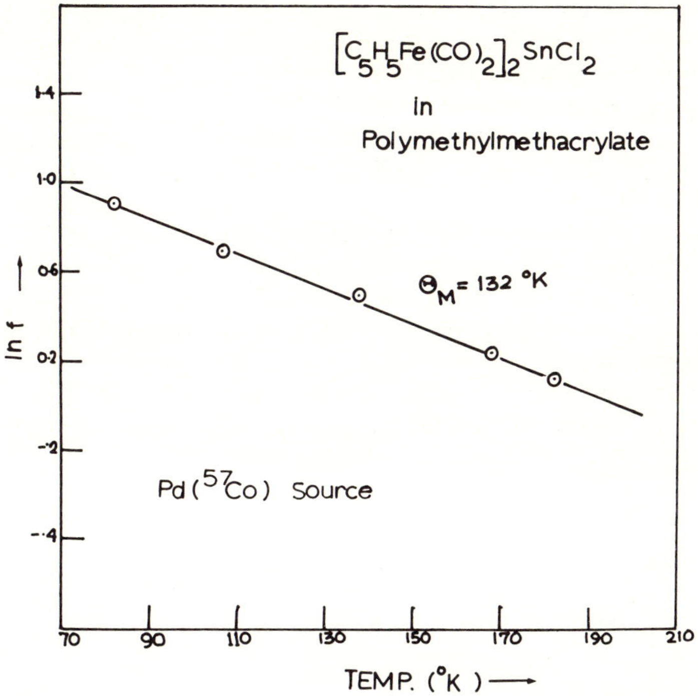

Fig. 7. Curves of ln *f* *vs.* *T* for $[C_5H_5Fe(CO)_2]_2SnCl_2$ in PMM Pd(Co^{57}) source.

the glassy matrix absorbers are, respectively, within the quoted experimental error of each other, as seen from the data summarized in Table I. Any significant *lengthening* of the Sn–Fe bond or *shortening* of the Sn–Cl bond would have been reflected in a change in the isomer shifts of the two resonant atoms in question. Similarly, any significant change in the Fe–Sn–Fe or Cl–Sn–Cl bond angles would be expected to be reflected in systematic changes in the quadrupole splitting parameters (see below). From the data of Table I it is seen that such changes have not occurred in going from the neat solid to the glassy matrix, and hence it is reasonable to conclude that the anomalous bond distances and bond angles which are observed in the X-ray diffraction data for the solids are a reflection of the bonding interactions within the molecule and are not ascribable to the intermolecular forces in the solid inherent in the "stacking" process by which the crystal is grown from solution.

A strikingly different phenomenon is observed in the Sn^{119} resonance of the bis(π-cyclopentadienyliron dicarbonyl)tin dichloride. The Sn^{119} resonance spectrum of the glassy solution matrix absorber (Fig. 6) shows the presence of two doublets, and these are associated with the existence of a pair of rotational isomers in the frozen solution. The two possible isomers differ from each other by a rotation of the Fe–Sn bond.

The Fe^{57} Mössbauer spectra for $[(\pi\text{-}C_5H_5)Fe(CO)_2]_2SnCl_2$ in the glassy matrix (Fig. 9), on the other hand, show that the local environment about the iron atom has remained nearly invariant on dissolution of the neat crystalline solid, consistent with the above interpretation.

It is appropriate to mention briefly the results of some other studies in which Mössbauer parameters have been obtained for

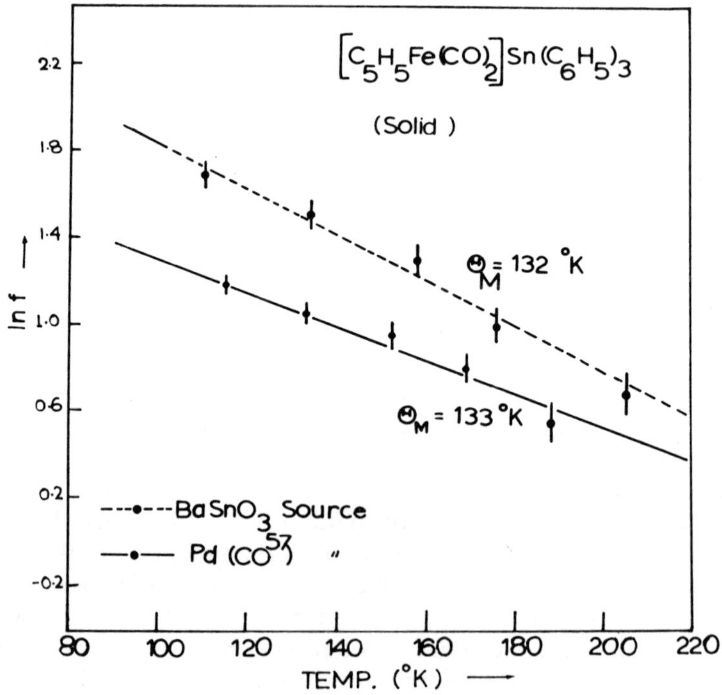

Fig. 8. ln f *vs.* T for $[C_5H_5Fe(CO)_2]Sn(C_6H_5)_3$ as neat solid: $BaSn^{119m}O_3$ and $Pd(Co^{57})$ sources.

frozen-solution systems in order to elucidate specific chemical problems which are susceptible to such a study. As part of a detailed study of alkyl tin cyanides and thiocyanates, Gassenheimer and Herber [22] have reported the results obtained for $(C_4H_9)_2Sn(SCN)_2$ under varying conditions. From the systematics for related compounds and the available infrared data it has been shown that such dialkyl tin dithiocyanates are hexacoordinate in the solid, with the two alkyl groups occupying *trans* octahedral positions and the SCN groups acting as bidentate bridging entities between adjacent tin atoms. Such a polymeric structure is consistent with the low solubility of this compound in nonpolar solvents, and its relatively high melting point. The considerable solubility of this compound in polar solvents raises the possibility of a conformational change on solution in such solvents, and a number of Mössbauer experiments have been carried out to clarify whether such structural changes do, in fact, take place.

Table III

Compound	State	*IS*, mm/sec	*QS*, mm/sec
$(C_4H_9)_2Sn(SCN)_2$	Neat solid	1.56 ± 0.05	3.88 ± 0.05
	(0.8*M* in MHF)	1.47 ± 0.05	3.98 ± 0.05
	(0.8*M* in CH_3CN)	1.49 ± 0.05	3.86 ± 0.05

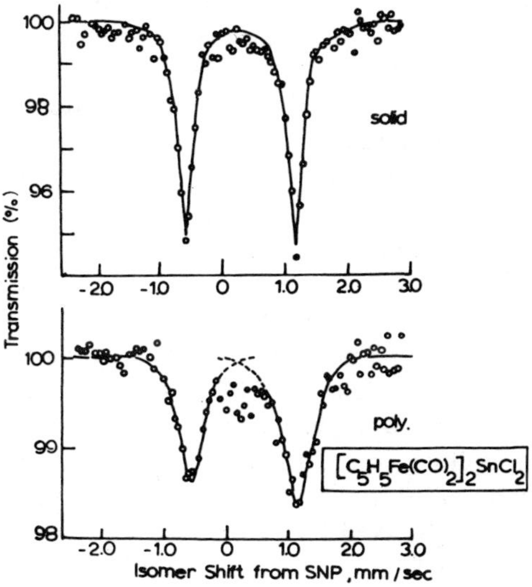

Fig. 9. Fe^{57} Mössbauer spectra of $[C_5H_5Fe(CO)_2]_2SnCl_2$ as neat solid and in PMM at 80°K. $Pd(Co^{57})$ source. The isomer shift is with respect to SNP at 294°K.

The Mössbauer parameters in $(C_4H_9)_2Sn(SCN)_2$ for the neat solid and the frozen acetonitrile and MTF solutions are found to be nearly identical (Table III). This may be readily understood in terms of the similarity of the bonding properties of the SCN bridging groups in $(C_4H_9)_2Sn(SCN)_2$ and the —C$\equiv$N: and 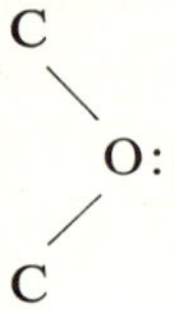 groups in CH_3CN and MTF, respectively. Furthermore, the observed similarity in the frozen-solution parameters for acetonitrile solutions (crystalline) and MTF solutions (glassy) lends added support to the assumption that the Mössbauer parameters reflect the nature of the metal–ligand bonding rather than structural features imposed by the nature of the surrounding matrix.

By comparing the results obtained with frozen solutions to those obtained with neat solids it has become possible to resolve some of the apparent disagreements concerning the changes in structure and conformation which occur when a solid dissolves in various solvents. Similar experiments along these lines have been carried out by a group of Israeli investigators [4]. Using the ME resonance in I_2^{119} they were able to study in detail the bonding of molecular iodine in the solid, in polar solvents (brown color), in nonpolar solvents (violet color), and in the presence of an adsorbent such as soluble starch (blue color). From such studies has come a much clearer understanding of the relationship between the color of iodine and its electronic configuration in different environments.

It is known, for example, that the maximum absorption of molecular iodine in the gas phase occurs at about 5200 Å, and that this characteristic absorption is also observed in iodine solutions in chloroform, carbon tetrachloride, carbon disulfide, and a number of hydrocarbons. On the other hand, in benzene this characteristic absorption is shifted to 5000 Å, in xylene to 4950 Å, and in acetone all the way to 4600 Å. The recent I^{129} ME experiments of Bukshpan *et al.* [4] have shown that both the isomer shifts and the quadrupole couplings found for I_2 in inert "solvents" such as hexane, carbon tetrachloride, and argon are very similar to each other and to the value observed for pure molecular I_2. In benzene solutions, however, they find the quadrupole coupling and the isomer shift considerably different, and have concluded that approximately 0.26 electrons are transferred from the benzene ring to the I_2 molecule. This observation confirms the earlier interpretation of the spectroscopic data for I_2 in benzene—as compared to inert solvents—as implying the formation of a benzene–I_2 complex in solution. There has been considerable speculation concerning the structures of such complexes, the two most likely structures being the one in which the I_2 axis is parallel to the benzene ring and the one in which the I_2 axis is perpendicular. Since the Mössbauer transition in I^{129} involves a transition between a 7/2 and a 5/2 spin state, it is possible to observe eight lines in the ME spectrum. Detailed analysis of the position and intensity of these lines permits the evaluation not only of the isomer shift and quadrupole splitting from such data, but, more importantly, from a single I^{129} spectrum one can calculate the sign of the electrostatic field gradient and the extent to which the symmetry of the electrostatic field is cylindrical with respect to the chemical bond axis.

It is this latter parameter which has been used to elucidate the probable structure of the I_2–benzene complex. The model in which the I_2 molecule lies parallel to the plane of the benzene ring predicts lack of spherical symmetry along the I_2 bond axis whereas the perpendicular model predicts spherical symmetry of the electrostatic field. The results of the ME experiments carried out by the Israeli group show very clearly that the asymmetry parameter is zero; that is, that there is cylindrical symmetry along the iodine–iodine bond axis. On this basis the model which envisions the I_2 molecule oriented perpendicularly with respect to the benzene ring appears to be the preferable structure.

ACKNOWLEDGMENT

The experimental work cited in this paper has been supported in part by the Petroleum Research Fund administered by the American Chemical Society, the Research Council of Rutgers University, and the US Atomic Energy Commission (Document NYO-2472-58). This support is gratefully acknowledged. The authors are also indebted to Dr. Y. Hazony for a number of fruitful discussions bearing on this work.

REFERENCES

1. V. Goldanskii and R. H. Herber (eds.), *Chemical Applications of Mössbauer Spectroscopy* (Academic Press, New York, 1968).

2. R. H. Herber in: *The Mössbauer Effect* (Symposium of the Faraday Society, No. 1, 1967) (Plenum Press, New York, 1968), pp. 86–96.
3. H. A. Stöckler, H. Sano, and R. H. Herber, *J. Am. Chem. Soc.* **89**:6864 (1967).
4. S. Bukshpan, C. Goldstein, and T. Sonino, *J. Chem. Phys.* **49**:5477 (1968).
5. A. J. Nozik and M. Kaplan, *J. Chem. Phys.* **47**:2960 (1967).
6. R. L. Mössbauer and W. H. Wiedemann, *Z. Physik* **159**:33 (1960).
7. A. H. Muir, Jr., "Tables and Graphs for Computing Debye–Waller Factors in Mössbauer Effect Studies," AI-6699 (Atomics International, California, 1962).
8. T. A. Manuel and F. G. A. Stone, *Proc. Chem. Soc.* p. 90 (1959); *J. Am. Chem. Soc.* **82**:336 (1960).
9. M. D. Rausch and G. N. Schrauzer, *Chem. Ind.* (*London*) **1959**:957.
10. A. Nakamura and N. Hagihara, *Bull. Chem. Soc. Japan* **32**:880.
11. F. A. Cotton, *J. Chem. Soc.* **1960**:400.
12. B. Dickens and W. N. Lipscomb, *J. Am. Chem. Soc.* **83**:489 (1961).
13. C. G. Kreiter, A. Maasbol, F. A. L. Anet, H. D. Kaesz, and S. Winstein, *J. Am. Chem. Soc.* **88**:3444 (1966).
14. C. E. Keller, B. A. Shoulders, and R. Pettit, *J. Am. Chem. Soc.* **88**:4760 (1966).
15. F. A. Cotton, A. Davison, and J. W. Faller, *J. Am. Chem. Soc.* **88**:4507 (1966).
16. R. Grubbs, R. Breslow, R. Herber, and S. J. Lippard, *J. Am. Chem. Soc.* **89**:6864 (1967).
17. R. T. Bailey, E. R. Lippincott, and D. Steele, *J. Am. Chem. Soc.* **87**:5346 (1965).
18. R. H. Herber and Y. Goscinny, *Inorg. Chem.* **7**:1293 (1968).
19. S. Bukshpan and R. H. Herber, *J. Chem. Phys.* **46**:3375 (1967).
20. Y. Hazony, *J. Chem. Phys.* **49**:159 (1968).
21. J. E. O'Connor and E. R. Corey, *Inorg. Chem.* **6**:968 (1967).
22. B. Gassenheimer and R. H. Herber, *Inorg. Chem.* **8**:1120 (1969).

Mössbauer Effect Studies on Eu^{151} in Mixed Oxide Structures

G. W. Dulaney* and A. F. Clifford

Department of Chemistry
Virginia Polytechnic Institute
Blacksburg, Virginia

In an effort to obtain more nearly monochromatic Sm^{151} *sources for* Eu^{151} *a study of the effect of crystal structure on absorption linewidth was undertaken. The broadest lines were found for ternary oxides of presumed high crystal symmetry, such as* $EuCrO_3$, *while the narrowest lines were found for the soft, asymmetric anhydrous* $EuCl_3$. *In the case of* $Eu_2Ti_2O_7$ *quadrupole splitting was observed corresponding to* 1.30 ± 0.04 *for the ratio of the quadrupole moments of the excited and ground states and with a value of* 1.00 ± 0.02 mm/sec *for* Γ. *The measured quadrupole coupling constant was* -280 ± 7 Mc/sec. *Absorption efficiencies were found, as expected, to be greatest for the most rigid crystals and least for the soft ones. Isomer shifts were found to vary in order of intuitive "ionicity," but in some cases the correlation was much better with mass than with electronegativity.*

INTRODUCTION AND EXPERIMENTAL

Past Mössbauer studies on europium, utilizing either Eu^{151} or Eu^{153}, have included studies of the magnetic hyperfine structure [1–3], the effect of composition on isomer shift [4–9], isomer-shift ratios [10, 11], and, recently, a study on the Mössbauer effect in europium chelates [12]. One of the problems which has beset the use of Mössbauer spectroscopy for studies on europium and a number of the other rare-earth elements is that the observed resonance lines are usually several times the natural linewidths. Since this results in a diminution in the resolvability of quadrupole-split spectra with concomitant loss of information, we have addressed ourselves to a search for the reasons for this line broadening and for a possible solution to the problem.

*Presently at the Digital Equipment Corporation, Maynard, Massachusetts.

The spectrometers used for this work were of the Kankeleit type operated at their resonant frequency in the multiscaling mode. Linearity was on the order of 0.1% and instability less than 0.1%. Spectra were normally taken with the source moving and with detection of the radiation by a gas-filled proportional tube. Most of the spectra reported were obtained with a xenon–nitrogen tube, although xenon–methane has been used, and currently we are using xenon–carbon dioxide. The spectrometers were calibrated regularly against sodium nitroprusside and iron foil.

Preparation of samples was by standard methods. The mixed oxides were made by grinding stoichiometric amounts of the constituent simple oxides together and heating the sample, contained in a platinum boat, at 900–1000°C for 10–36 hr. Structures were verified by X-ray diffraction and any excess Eu_2O_3 was leached out by boiling with a saturated solution of NH_4Cl for 3–4 hr.

The fluorides were generally made by precipitating from aqueous solution, drying at 110°C in vacuum, and then dehydrating in SF_4 at 250–400°C for $1\frac{1}{2}$–3 hr.

Anhydrous $EuCl_3$ was prepared by suspending the sample over refluxing $SOCl_2$ for 30 hr, keeping the sample at 200–300°C.

The sulfates were prepared by crystallizing from aqueous solution, the double cesium sulfate being derived from an equimolar mixture of Cs_2SO_4 and $Eu_2(SO_4)_3$.

The radiation source in all of these studies was anhydrous $Sm^{151}F_3$. This was prepared as described above. A typical spectrum of EuF_3, in this case prepared by heating a mixture of $EuF_3 \cdot nH_2O$ with excess NH_4HF_2 until all the NH_4HF_2 was sublimed, shown in Fig. 1, had $IS = 0.007 \pm 0.006$ mm/sec, possibly indicating a slightly different electronic environment in the two fluorides. Contrary to what has been reported elsewhere (e.g., ref. [9]), however, the isomer shift of the oxide, Eu_2O_3, was found to be quite different from that of the fluoride, with Eu_2O_3 (cubic, *C*-type) absorbing at 1.037 ± 0.014 mm/sec and Eu_2O_3 (monoclinic) at 1.030 ± 0.007 mm/sec relative to the SmF_3 source. Since the trifluoride precipitated from aqueous solution readily hydrolyzes when heated in moist air, it seems very likely that some studies have been done on the oxide rather than the fluoride as reported. The resonance for $EuSO_4$ (Fig. 2) occurred at -13.82 ± 0.02 mm/sec relative to the fluoride or -14.86 ± 0.03 relative to Eu_2O_3. This compares with -15.0 mm/sec reported by Hüfner *et al.* [7]. Similarly, we observe the resonance for $Eu_2(SO_4)_3 \cdot 8H_2O$

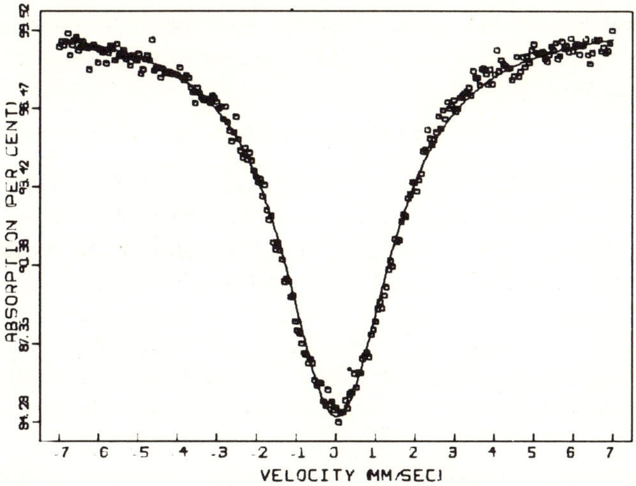

Fig. 1. Mössbauer spectrum of EuF_3 (per cent absorption: 15.5; Γ: 1.66 ± 0.02 mm/sec; *IS*: 0.007 ± 0.007 mm/sec).

at 0.221 ± 0.008 mm/sec *vs.* EuF_3 or −0.816 *vs.* Eu_2O_3 compared to −0.79 ± 0.04 mm/sec for "$Eu_2(SO_4)_3$" as reported by Deeney *et al.* [9]. On the other hand, we observed the resonance for $EuCl_3 \cdot {\sim}3.5H_2O$ at 0.341 mm, or −0.696 mm/sec *vs.* Eu_2O_3, while Deeney reports it at −0.42 ± 0.04 for "$EuCl_3$." We can only conclude

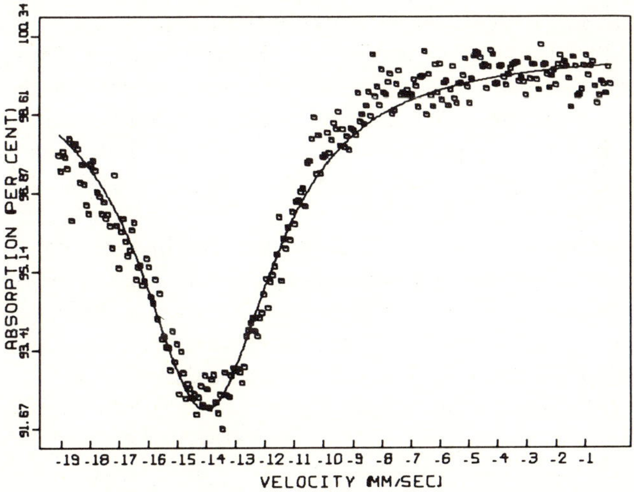

Fig. 2. Mössbauer spectrum of $EuSO_4$ (per cent absorption: 7.9; Γ: 2.79 ± 0.04 mm/sec; *IS*: −14.03 ± 0.02 mm/sec).

that Deeney's chloride was extensively hydrolyzed. It consequently must be emphasized that proper sample preparation and identification are of great importance.

LINEWIDTHS

The width (full-width at half-maximum, FWHM) of all of these resonances is considerably greater than the "natural" width of 1.33 mm/sec, varying from 4.46 ± 0.16 mm/sec for $EuSO_4$, through 3.40 ± 0.08 mm/sec for cubic Eu_2O_3, to lower values for some other compounds.

It seems reasonable to believe that much of the line broadening observed is the result of incipient splitting, and that, since many of the lines are demonstrably asymmetrical, an important factor may be quadrupole splitting. Since most rare-earth compounds are of low crystallographic symmetry and since for other elements quadrupole splitting is generally eliminated when the element in question is in a site of high symmetry, it was decided to study the effect of crystal structure on the linewidth of europium absorbers. A search was therefore undertaken for compounds of high crystallographic symmetry with the expectation that they would be characterized by narrower resonance lines.

In the series of compounds $EuMO_3$, which, when M is aluminum or a first transition-series trivalent ion, have perovskite structure, the rare-earth ion is probably in a highly symmetrical 12-coordinate site and should therefore give rise to no quadrupole splitting. Consequently, expecting narrower linewidths, we prepared several of these compounds in which M = Al, Cr, Fe, or Co, and were surprised to find that these gave exceptionally broad resonance lines amounting to FWHM = 3.57 ± 0.03 mm/sec in the case of $EuCoO_3$ (Fig. 3) and 4.01 (±0.04) mm/sec in $EuCrO_3$. It therefore appeared that increasing the symmetry actually resulted in greater line broadening.

A second series of compounds has also been examined, in which the crystallographic symmetry is low. In one of these, anhydrous $EuCl_3$ (Fig. 4), we have found the narrowest resonance line thus far observed, FWHM = 2.39 ± 0.02 mm/sec, despite the fact that this substance has the weakest crystal forces of any substance examined by us except for anhydrous $EuBr_3$ (FWHM = 2.42 ± 0.02 mm/sec). Other narrow linewidths have been found in $Eu_2(SO_4)_3 \cdot 8H_2O$ (FWHM = 2.75 ± 0.10 mm/sec), in $CsEu(SO_4)_2 \cdot 12H_2O$ (FWHM = 2.60 ± 0.10 mm/sec), and in anhydrous $Eu_2(SO_4)_3$ (FWHM = 2.56

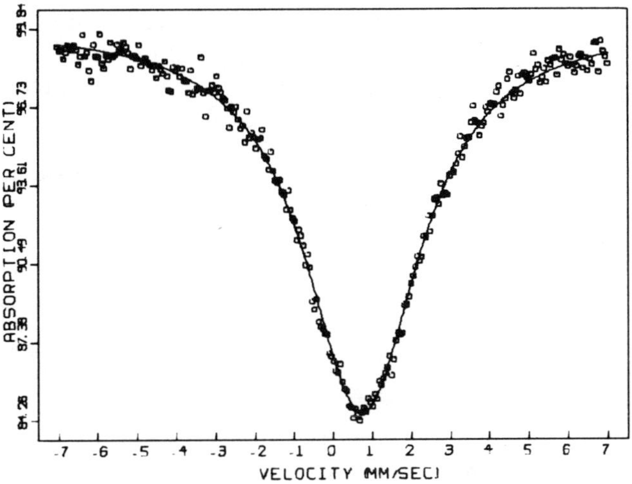

Fig. 3. Mössbauer spectrum of $EuCoO_3$ (per cent absorption: 15.5; Γ: 1.79 ± 0.02 mm/sec; *IS*: 0.727 ± 0.008 mm/sec).

mm/sec). In these the environment of the europium atoms is highly asymmetrical, with a complex arrangement of sulfate ions and water molecules around the europium atoms. The samarium analog, $Sm_2(SO_4)_3 \cdot 8H_2O$, apparently has very low symmetry, the crystal

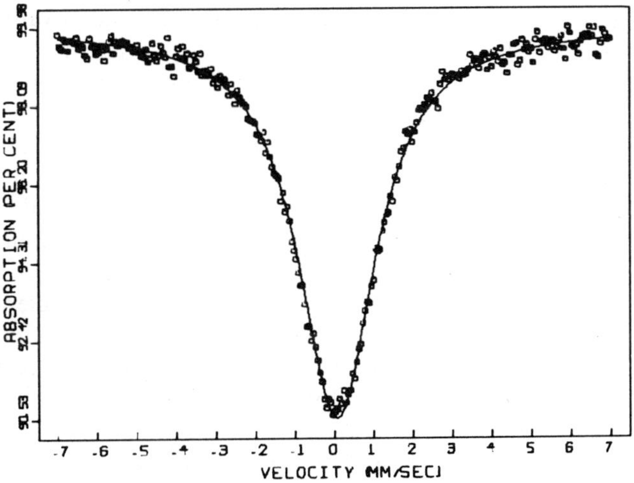

Fig. 4. Mössbauer spectrum of anhydrous $EuCl_3$ (per cent absorption: 9.4; Γ: 1.19 ± 0.01 mm/sec; *IS*: 0.049 ± 0.005 mm/sec).

structure [13] being monoclinic, space group C_{2h}^6. We are thus faced with the apparent paradox that in europium compounds the more crystallographically symmetrical compounds exhibit the greatest line broadening. The data are summarized in Table I.

Table I. Linewidths (FWHM) (in mm/sec) of Europium Compounds at Room Temperature*

$EuSO_4$	4.45 ± 0.08	EuF_3†	3.32 ± 0.04
$EuBO_3$	4.13 ± 0.04	Eu_2FeNbO_6‡	3.30 ± 0.08
$EuCrO_3$	4.01 ± 0.04	$NaEuF_4$	3.30 ± 0.04
EuOF	3.70 ± 0.10	EuF_3§	2.92 ± 0.06
$EuCoO_3$	3.57 ± 0.03	Eu_2FeNbO_6	2.89 ± 0.07
$EuAlO_3$	3.52 ± 0.04	BiF_3 (10% Eu)	2.82 ± 0.04
$EuF_3 \cdot nH_2O$	3.52 ± 0.06	$Eu_2(SO_4)_3 \cdot 8H_2O$	2.75 ± 0.10
$EuCl_2$	3.51 ± 0.07	$CsEu(SO_4)_2 \cdot 12H_2O$	2.60 ± 0.06
Eu_2O_3 (monoclinic)	3.46 ± 0.02	$Eu_2(SO_4)_3$	2.56 ± 0.02
$EuFeO_3$	3.46 ± 0.05	$EuCl_3 \sim 3.5H_2O$	2.54 ± 0.10
Eu_2O_3 (cubic)	3.40 ± 0.08	$EuBr_3$	2.42 ± 0.04
Ba_2EuNbO_6	3.34 ± 0.05	$EuCl_3$	2.39 ± 0.02
Eu_3NbO_7	3.34 ± 0.05		

*All samples 20–30 mg/cm^2.
†Prepared with NH_4HF_2.
‡95°K.
§Prepared with SF_4.

It can be seen that of the trivalent substances, those exhibiting the greatest linewidths (4.1–2.9 mm/sec) are those with the most-rigid lattices. Of intermediate linewidth (2.8–2.5 mm/sec) are the salts with discrete, nonpolarizable anions (SO_4^{2-}, F^-) or that are hydrated. The lowest linewidths are exhibited by the anhydrous chloride and bromide, which have very low crystal forces in comparison to the others.

That the line broadening is not the result of loss to crystal vibration is shown by the fact that with the exception of the compounds containing other heavy elements like barium and niobium, which predictably decrease resonant absorption, the absorption efficiency is, as expected, highest for the substances with the most-rigid crystals. A tabulation is given in Table II. In the light of this evidence it seems necessary to attribute the line broadening to an effect of crystal structure.

A qualitative explanation of the broadening in terms of quadrupole splitting may be as follows: The seven occupied f orbitals of the europium atom cannot be oriented symmetrically in space, or in such

Table II. Absorption Efficiencies (%) of Europium Compounds at Room Temperature*

$EuBO_3$	17.7	$EuOF$	11
$EuCrO_3$	16.4	Eu_2FeNbO_6	10.6
$EuCoO_3$	15.5	$EuCl_3$	9.4
EuF_3†	15.5	Eu_3NbO_7	9.2
$EuAlO_3$	15	$Eu_2(SO_4)_3$	9.2
$Eu_2Ti_2O_7$	14.5	Eu_2FeNbO_6‡	8.8
$EuFeO_3$	12.8	$CsEu(SO_4)_2 \cdot 12H_2O$	8.6
$NaEuF_4$	12.8	$Eu_2(SO_4)_3 \cdot 8H_2O$	8
$EuF_3 \cdot nH_2O$	12	$EuBr_3$	6.2
Eu_2O_3 (cubic)	12	Ba_2EuNbO_6	6
Eu_2O_3 (monoclinic)	11.4	$EuCl_3 \cdot \sim 3.5H_2O$	4.5

*All samples 20–30 mg/cm^2.
†Prepared with NH_4HF_2.
‡95°K.

a way that they would direct equal electron density toward the lattice sites in a rigid, highly symmetrical crystal, whose symmetry is dictated by geometric factors. Consequently, the field of the *f* electrons is distorted by the surrounding atoms in a highly symmetrical crystal such as $Eu_2Ti_2O_7$ or $EuCrO_3$, giving rise to a nonzero electric field gradient at the europium nucleus and consequent quadrupole splitting. In a soft crystal such as $EuCl_3$ or $CsEu(SO_4)_2 \cdot 12H_2O$, in which the europium atom is surrounded by water molecules and is closer to the state of the free ion, the positions of the ligand atoms or molecules are dictated by the orientation of the *f* orbitals and are located in such a way as to maximize the symmetry of *f*-electron distribution. This results in minimum field gradient and narrower lines. It may be significant in this regard that most rare-earth compounds, especially those having less-rigid crystals, have very low symmetry, e.g., the hydrated salts, anhydrous halides, double sulfates, and nitrates and many more. It will be an interesting problem to calculate the electron distributions in these systems.

A test of this theory would be the observation of quadrupole splitting in a compound having europium in a site of high symmetry. We believe we have found such a case in $Eu_2Ti_2O_7$, which has cubic pyrochlore structure and which exhibits obvious quadrupole splitting consistent with theory (Fig. 5). This compound was shown by X-ray studies to be crystallographically pure. It is interesting to note that the linewidth in this spectrum, taken without regard for the splitting, is

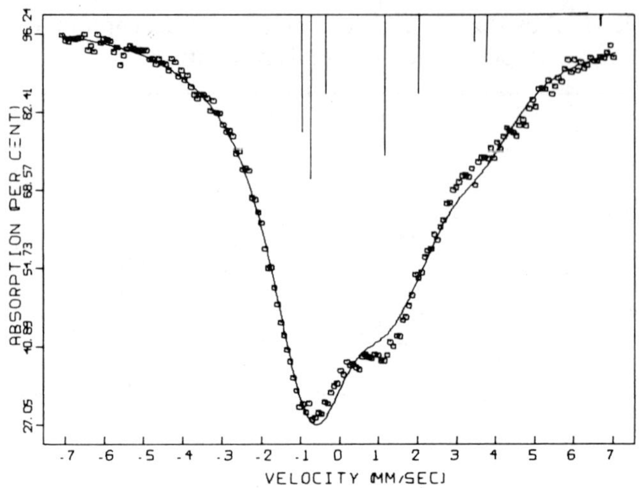

Fig. 5. Mössbauer spectrum of $EuTiO_{3.5}$ with theoretical lines (maximum per cent absorption: 8.4; Γ: 1.15 $\pm$ 0.15 mm/sec; *IS*: 0.58 $\pm$ 0.03 mm/sec; eqQ_{gd}: -278 ± 5 Mc/sec; Q_{ex}/Q_{gd}: 1.313 $\pm$ 0.014).

close to 4 mm/sec, comparable with the largest widths listed in Table I. On the other hand, the computed width of the most intense peak is only 2.2 mm/sec, lower even than that for $EuCl_3$. From this spectrum we obtain a quadrupole coupling constant of -260 mc/sec and a ratio of quadrupole moments of the excited state to the ground state of 1.31, which is in agreement with Nowik and Ofer's [2] value of 1.3 $\pm$ 0.5.

Recoilless Fraction and Natural Linewidth

As part of the general investigation of linewidths, the width of the EuF_3 line was studied as a function of the density of the sample. The results of this study are shown in Fig. 6 and Table III and lead to a natural linewidth of 1.33 mm/sec and a recoilless fraction of 0.4 $\pm$ 0.05. This latter is in line with that of other rare earths.

The same effect has been observed for samples of $NaEuF_4$ of varying density.

ISOMER SHIFTS

As a result of this study a considerable number of isomer shifts for Eu^{151} have been added to those already in the literature [3–9]. These are shown in Table IV and appear roughly to reflect the changing

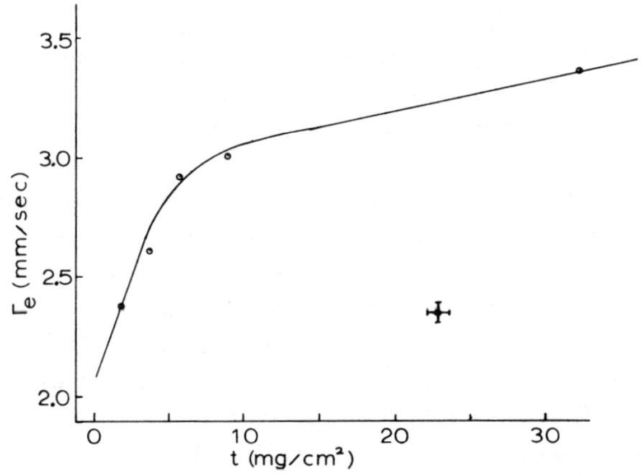

Fig. 6. Experimental linewidth *vs.* absorber surface density for EuF_3.

Table III. Effect of Density on Linewidth

Density, mg/cm² ± 10%	Γ_{expt},* mm/sec, ± 1%
1.90	2.38
3.83	2.64
5.88	2.92
9.05	3.05
32.4	3.32

*$\Gamma_{expt} = Df' + \Gamma_{abs}{}^0 + \Gamma_{source}{}^0$, where $D = 0.27\,\Gamma_{nat} n\sigma_0 t$, f' is the recoilless fraction, $\sigma_0 \equiv 5.2 \times 10^{-19}$ cm², and $\Gamma_{nat} \equiv 1.33$ mm/sec.

Table IV. Effect of Ionicity on Isomer Shifts (in mm/sec) *vs.* EuF_3*

Eu_2O_3 (cubic)	1.037 ± 0.014	BiF_3 (10% Eu)	0.210 ± 0.010
Eu_2O_3 (monoclinic)	1.030 ± 0.007	EuOF	0.171 ± 0.014
Ba_2EuNbO_6	0.991 ± 0.015	$NaEuF_4$	0.170 ± 0.004
Eu_3NbO_7	0.895 ± 0.015	$EuBr_3$	0.166 ± 0.007
Eu_2FeNbO_6†	0.874 ± 0.012	$CsEu(SO_4)_2 \cdot 12H_2O$	0.101 ± 0.017
Eu_2FeNbO_6	0.788 ± 0.014	$EuF_3 \cdot nH_2O$	0.053 ± 0.013
$EuCrO_3$	0.746 ± 0.007	$EuCl_3$	0.049 ± 0.005
$EuFeO_3$	0.733 ± 0.015	$Eu_2(SO_4)_3$	0.009 ± 0.008
$EuCoO_3$	0.727 ± 0.006	EuF_3‡	0.007 ± 0.006
$EuAlO_3$	0.586 ± 0.007	EuF_3	0.000 ± 0.010
$EuBO_3$	0.537 ± 0.008	$EuCl_2$	−13.103 ± 0.050
$EuCl_3 \cdot \sim 3.5H_2O$	0.341 ± 0.007	$EuSO_4$	−13.82 ± 0.02
$Eu_2(SO_4)_3 \cdot 8H_2O$	0.221 ± 0.008		

*All samples 20–30 mg/cm².
†95°K.
‡Prepared with NH_4HF_2.

degree of ionicity. The series is entirely consistent with that given by Deeney *et al.* [9] when correction is made for the position of EuF_3 in the series. Deeney's value for $Eu(ClO_4)_3$ ($IS = 0.23$ mm/sec *vs.* EuF_3) is quite consistent with a trend of increasing ionicity going from Eu_2O_3 to EuF_3, as is also the *IS* for the acetate (0.61 *vs.* EuF_3). The values for the chloride and sulfate have already been commented on.

Note should be taken of the great difference in isomer shift, and indeed also of the linewidths, between the hydrated and anhydrous forms of EuF_3, $Eu_2(SO_4)_3$, and $EuCl_3$, once again pointing out the importance of identifying and reporting the exact forms under study.

Attempts have been quite successful in many cases to correlate the isomer shift with the electronegativity of the element associated with the Mössbauer element e.g., [14]). In the case of very large differences of electronegativities, however, the simple relationship between electronegativity or electronegativity difference and the property being measured no longer applies, because the bonding is no longer predominantly covalent [15]. An attempt to correlate the isomer shift with the electronegativity of the halogen in the trivalent halides (Fig. 7) shows no simple relationship, as might have been expected. It is also obvious that no simple relationship exists for the series of compounds $EuMO_3$ as a function of the electronegativity of M. On the other hand, if the isomer shift is correlated with the mass of the halide ion, or with the mass of M, then, as can be seen in Fig. 8, a good linear relationship is found. Isomer shifts that appear to be

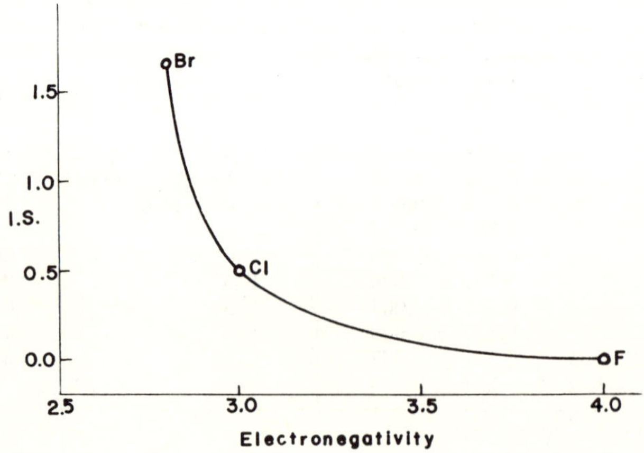

Fig. 7. Mössbauer isomer shift *vs.* electronegativity for some europium trihalides.

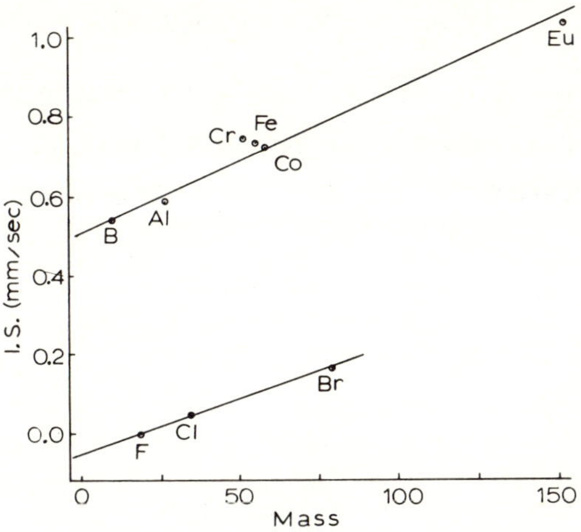

Fig. 8. Mössbauer isomer shift *vs.* mass of M or X in $EuMO_3$ and EuX_3.

dependent on mass have been observed before (e.g., [16]). A logical explanation for this invokes the second-order Doppler effect, which, in concert with the changes in force constants required by the changes in the nature of the crystals, could produce the observed effect.

In summary, it is obvious that much yet remains to be done both in the technology and in the interpretation of rare-earth Mössbauer spectroscopy. It is hoped that the present studies will help to provide solutions for some of these problems.

ACKNOWLEDGMENT

We gratefully acknowledge the support of the Atomic Energy Commission through its Grant No. AT(40-1)-3603.

REFERENCES

1. P. H. Barrett and D. A. Shirley, *Phys. Rev.* **131**:123 (1963).
2. I. Nowik and S. Ofer, *Phys. Rev.* **132**:241 (1963).
3. H. H. Wickman, I. Nowik, J. H. Wernick, D. A. Shirley, and R. B. Frankel, *J. Appl. Phys.* **37**:1246 (1966).
4. P. Brix, S. Hüfner, P. Kienle, and D. Quitmann, *Physics Letters* **13**:140 (1964).
5. I. Nowik and S. Ofer, *Rev. Mod. Phys.* **36**:392 (1964).
6. E. Steichele, S. Hüfner, and P. Kienle, *Physics Letters* **14**:321 (1965).
7. S. Hüfner, P. Kienle, D. Quitmann, and P. Brix, *Z. Phys.* **187**:67 (1965).
8. U. Atzmony, E. R. Bauminger, I. Nowik, S. Ofer, and J. H. Wernick, *Phys. Rev.* **156**:263 (1967).
9. F. A. Deeney, J. A. Delaney, and V. P. Ruddy, *Physics Letters* **25A**:370 (1967).
10. U. Atzmony and S. Ofer, *Physics Letters* **14**:284 (1965).
11. E. Steichele, S. Hüfner, and P. Kienle, *Physics Letters* **21**:220 (1966).

12. F. A. Deeney, J. A. Delaney, and V. P. Ruddy, *J. Inorg. Nucl. Chem.* **30**:1175 (1968).
13. W. H. Zachariasen, *J. Chem. Phys.* **3**:197 (1935).
14. M. Cordey-Hayes, *J. Inorg. Nucl. Chem.* **26**:915 (1964); D A. Shirley, *Rev. Mod. Phys.* **36**: 339 (1964).
15. A. F. Clifford, *J. Am. Chem. Soc.* **79**:5404 (1957); *J. Phys. Chem.* **63**:1227 (1959).
16. A. J. F. Boyle and G. J. Perlow, *Bull. Am. Phys. Soc.* **10**:482 (1965).

Systematic Interpretation of the Isomer Shifts in Tin, Antimony, Tellurium, Iodine, and Xenon*

G. K. Shenoy and S. L. Ruby

Argonne National Laboratory
Argonne, Illinois

The experimental isomer shifts for compounds of tin, antimony, tellurium, iodine and xenon have been compared with atomic Hartree–Fock density calculations for the chemically appropriate electronic configurations. This leads to values of $\delta\langle R^2\rangle/\langle R^2\rangle$ *for each nucleus. In addition, the isomer shifts for isoelectronic pairs of compounds (e.g.,* SnO *with* Sb_2O_3, Sb_2O_3 *with* TeO_2, *etc.) have been compared to find ratios of* $\delta\langle R^2\rangle/\langle R^2\rangle$ *for adjacent nuclei. These ratios agree closely with those deduced from the first method. This agreement gives support to the simple chemical interpretation of the isomer shifts for this series of nuclei and at the same time supplies a technique for deducing electronic configurations from the measured shifts.*

INTRODUCTION

The isomer shifts obtainable from nuclear gamma-ray resonance (NGR) measurements are of considerable importance in both chemical and nuclear investigations. To obtain the change in the mean-squared radius of a nucleus as it deexcites to the ground state, one must have detailed knowledge of the electron density at the nucleus in various chemical compounds, in addition to the experimental values of the isomer shifts.

The values of electron density at the nucleus in various electron configurations have mainly been estimated either by use of the Fermi–Segré relation or from the Hartree–Fock self-consistent-field (HF SCF) computations for atoms. However, the real electron configurations

*Work performed under the auspices of the US Atomic Energy Commission.

in solids may be different from the atomic description adopted in these computations. Hence, the deduced values of $\delta\langle R^2\rangle/\langle R^2\rangle$ are rather uncertain.

More-precise values can be found for the ratios of the change in nuclear-charge radii in two situations: (1) for different excited states of the same nucleus, the examples being Eu^{153}, Yb^{171}, Ir^{191}, Ir^{193}, etc., and (2) for different isotopes of the same element, for instance, in I^{127} and I^{129}, Ir^{191} and Ir^{193}, Yb^{170} and Yb^{171}, Gd^{155} and Gd^{156}, etc.

These approaches are completely independent of the values of electron density. However, such a comparison is possible only in a few fortuitous cases [1]. A way to extend this approach was put forward first by Ruby *et al.* [2], who found the ratio $\delta\langle R^2\rangle/\langle R^2\rangle$ for Sn^{119} and Sb^{121} by comparing isoelectronic compounds of these two nuclei. As we shall discuss below, even though the method is not completely independent of the electron-density values, the results are not sensitive to computational technique. We have now extended this approach with fair success to determine sequentially the ratio of $\delta\langle R^2\rangle/\langle R^2\rangle$ for Sn^{119}, Sb^{121}, Te^{125}, I^{127}, I^{129} and Xe^{129}. This forms the major subject of this paper.

Since the early 1960's, numerous authors have developed computational techniques to obtain atomic wave functions in the nonrelativistic approximations [3]. Some work has also been done more recently to take account of relativistic effects [4]. In this work we have used two programs to compute the *s*-electron density for a variety of electronic configurations of the above-mentioned elements. The first one, developed by Froese and Wilson,* does detailed nonrelativistic HF computations in the SCF approximation while taking into account exchange along with other orthogonality restrictions. The second technique is the one published by Herman and Skillman [3]; it deals with exchange through Slater's approximation, but is again nonrelativistic. The densities obtained by the two techniques have been compared. These densities are nonrelativistic, and to transform to relativistic densities, we have used the relativity factor $S'(Z)$ as outlined by Shirley [6].

By assigning an electronic configuration to a particular solid-state absorber, the isomer shifts and the calculated densities can be used to deduce $\delta\langle R^2\rangle/\langle R^2\rangle$. These numbers, individually obtained for each of the above-mentioned isotopes, along with the ratios between them found earlier, are shown to be internally consistent.

*Dr. Michael Wilson developed the version of the Froese program [5] that we used.

EXPERIMENTAL RESULTS

Almost all the data reported here are taken from earlier work already in the literature. In a few instances it has seemed advisable to us to make some additional measurements. The data used for each element have been analyzed critically, and will be discussed for each element separately.

Sn^{119}

The 23.9-keV radiation from this nucleus has been studied extensively by use of the NGR technique. In particular, the absorbers employed have represented a wide variety of chemical environments. The discussion here is restricted to some of the simpler inorganic or metallic crystals, and the extensive results on organometallic compounds will not be used. The results for Sn (and Sb) differ in no important way from our earlier work [2], which can be referred to for the past history.

However, the numerous earlier measurements of isomeric shifts have not been mutually consistent. The earlier paper [2] attempted to refer all the tin shifts to α-tin, although no sample of α-tin was available for measurements to check the values from the literature. In addition, reports of previous authors were occasionally inconsistent with each other.

To clarify the situation, some of the more important inorganic compounds were rerun. In particular, it was desirable to check α-tin. To prepare the sample, pure white tin containing grains of gray tin was rolled into a 10 mg/cm^2 foil, folded double, rerolled to 10 mg/cm^2, and finally held at 233°K for a few hours. At the end of this time the material had completely transformed into α-tin.*

The new results indicate that systematic errors of 0.2 mm/sec had crept into the table of ref. [2]. Our data are now quite consistent with the values of Lees and Flinn [7]. The change of experimental values does not affect either $\delta\langle R^2\rangle/\langle R^2\rangle$ for Sn^{119} or the ratio of this quantity to that of Sb^{121} reported earlier.

Sb^{121}

A few new isomer shifts of Sb compounds [8] are reported in Table I, even though some of them will not be used as members of

*We wish to thank Prof. A. W. Ewald of Northwestern University for both advice and starting material to make this transformation.

Table I. Isomer Shifts for Antimony Compounds Referred to InSb

Compound	Shift, mm/sec	Ref.
$KSbF_6$	$+12.3 \pm 0.4$	[2]
$NaSbF_6$	$+10.4 \pm 0.2$	[9]
Sb_2O_5	$+9.9 \pm 0.2$	[2]
	$+8.4 \pm 0.2$	[9]
$NaSb(OH)_6$	$+8.9 \pm 0.2$	[9]
$HSbCl_6 \cdot xH_2O$	$+5.4 \pm 0.2$	[9]
$SbCl_5$	$+4.9 \pm 0.3$	[9]
InSb	0	
Sb in β-tin	-2.66 ± 0.1	[2]
Sb (metal)	-2.8 ± 0.2	[9]
Sb_2O_3	-3.02 ± 0.05	[2]
	-2.0 ± 0.3	[9]
SbF_3	-6.0 ± 0.2	[2]
SbI_3	-7.40 ± 0.10	[8]
$SbCl_3$	-5.30 ± 0.16	[8]

isoelectronic pairs. In addition to our work, that of a Russian group [9] has been reported. Their shifts are systematically somewhat less than ours, and the discrepancy is not understood.

Te^{125}

Until recently the reported isomer shifts of tellurium compounds had large error limits, the basic problem in the precise determination of shifts being the broadness of the observed lines and the smallness of the Mössbauer absorption. These difficulties have been considerably reduced by using I^{125} in a Cu matrix as source [10]. The recent measurements of Jung and Triftshauser [11] and of Ruby and Shenoy [8] have provided better statistics than earlier work, and are given in Table II, along with some of the earlier measurements [12, 13].

I^{127} and I^{129}

Most measurements on the I^{127} isotope come from Perlow and his group [14, 16]; the others listed are from Jung and Triftshauser [11] and Reddy *et al.* [17]. For consistency, the measurements have been reported relative to I^-.

Table II. Isomer Shifts of Tellurium Compounds Relative to I^{125} in Cu Source

Compound	Shift, mm/sec	Ref.
$(TeI_6)^{--}$	2.0 ± 0.3	[9]
$TeCl_4$	1.2 ± 0.1	[11]
	1.1 ± 0.1	[8]
TeI_4	1.0 ± 0.2	[8]
	1.8 ± 0.9	[11]
TeO_2	0.8 ± 0.1	[10]
	0.7 ± 0.1	[8]
	0.71 ± 0.16	[13]
	0.72 ± 0.07	[11]
Te	0.5 ± 0.1	[10]
	0.67 ± 0.03	[8]
	0.50 ± 0.16	[13]
	0.78 ± 0.10	[11]
$TeCl_2$	0.48 ± 0.20	[11]
	0.50 ± 0.20	[8]
TeF_4	0.40 ± 0.1	[12]
SnTe	0.28 ± 0.04	[8]
Na_2TeO_3	0.40 ± 0.1	[12]
	0.22 ± 0.05	[11]
K_2TeO_3	0.14 ± 0.07	[8]
ZnTe	0.10 ± 0.1	[10]
	0.10 ± 0.05	[8]
	−0.16 ± 0.04	[11]
Cu(I)	(0)	—
Na_2TeO_4	−0.99 ± 0.03	[11]
	−1.00 ± 0.1	[8]
K_2TeO_4	−0.98 ± 0.05	[11]
H_6TeO_6	−1.25 ± 0.04	[13]
	−1.15 ± 0.04	[11]
TeO_3	−1.07 ± 0.05	[11]
TeF_6	−1.40 ± 0.13	[8]

Almost all the contributions to the spectroscopy of I^{129} have come from Hafemeister *et al.* [18, 19] and from the Soreq group [20–23]. The shifts have been referred to a ZnTe source. The measurements relevant to the present work are given in Table III.

Xe^{129}

All the work on Xe^{129} compounds has been done by Perlow [14], from whom the measurements in Table IV were obtained. The shifts have been referred to Xe° absorber.

Table III. Isomer Shifts for Iodine Compounds*

Compound	Shifts for I^{127} referred to [I^-]	Shifts for I^{129} referred to [ZnTe]
[H_6TeO_6]	0.90 ± 0.02 [14]	−2.45 + 0.05 [20]
$Na_3H_2IO_6$	0.88 ± 0.01 [14] 1.19 ± 0.05 [17] 0.86 ± 0.04 [11]	−3.35 ± 0.02 [17]
KIO_4	0.68 ± 0.09 [17] 0.54 ± 0.02 [16]	−2.3 ± 0.06 [18, 19] −2.05 ± 0.05 [17]
$NaIO_4$	0.69 ± 0.07 [11]	—
I^-	(0)	—
KI	−0.2 ± 0.02 [14]	−0.51 ± 0.025 [18] −0.45 ± 0.03 [17]
NaI	−0.2 ± 0.02 [14]	−0.46 ± 0.025 [18]
CsI	−0.4 ± 0.02 [14]	−0.37 ± 0.025 [18]
[ZnTe]	−0.163 ± 0.02 [14]	(0)
I_2	−0.74 ± 0.04 [14]	0.83 ± 0.01 [21]
$KICl_2$	−0.74 ± 0.04 [16]	—
ICl	−0.78 ± 0.04 [14]	1.73 ± 0.05 [22]
$KICl_4$	−1.55 ± 0.05 [14, 16]	—
[$Te(NO_3)_4$]	—	2.52 ± 0.01 [23]
I_2Cl_6	—	3.50 ± 0.02 [22]

*The reference to the work is given after each value.
†The square brackets around the name of any compound indicate that it was the source in the measurement.

Table IV. Isomer Shifts for Xe^{129} Compounds Relative to Xe°

Compound	Shift,* mm/sec
XeF_4	0.40 ± 0.04
[$XeCl_4$]	0.25 ± 0.08
[$XeCl_2$]	0.17 ± 0.08
XeF_2	0.10 ± 0.12
Xe°	(0)
Na_4XeO_6	−0.19 ± 0.02
[XeO_4]	−0.22 ± 0.02
[XeO_6]$^{4-}$	−0.22 ± 0.02

*All results are from Perlow [14].

RATIO OF CHANGES IN NUCLEAR-CHARGE RADII

The theory of isomer shifts has been discussed at length by Shirley [6]. The shift can be expressed as

$$S = (c/E_\gamma)F(Z,A)\,\Delta\psi^2\rho \tag{1}$$

where $F(Z,A) = (2/5)(\pi Ze^2R^2)S'(Z)$ and $\rho = \delta\langle R^2\rangle/\langle R^2\rangle$. The symbols used are the usual ones. The $\Delta\psi^2$ expresses the change in non-relativistic *s*-electron density at the nucleus when the atom changes from one chemical state to another. The correction for the relativistic effects come through the Racah parameter $S'(Z)$. Had the computation of $\Delta\psi^2$ by use of the Dirac Hamiltonian been started *ab initio*, $S'(Z)$ would not have appeared. In this connection, we call the attention of the reader to useful remarks made in earlier references [2, 6].

The ratio of the values of ρ for two nuclei can now be expressed as

$$\frac{\rho_2}{\rho_1} = \frac{E_{\gamma_2}}{E_{\gamma_1}}\frac{F(Z_1,A_1)}{F(Z_2,A_2)}\frac{\Delta\psi_{Z_1}{}^2}{\Delta\psi_{Z_2}{}^2}\frac{S_2}{S_1} \tag{2}$$

where the subscripts refer to the two nuclei under consideration. This expression is quite as general as Eq. (1). Further, the expression demands a detailed knowledge of the change in electron density between the two chemical compounds in question for each of the two nuclei. In the special cases mentioned earlier, however, this expression has simpler meaning. First, if we are referring to two excited states of the same nucleus, we have

$$\frac{\rho_2}{\rho_1} = \frac{E_{\gamma_2}}{E_{\gamma_1}}\frac{S_2}{S_1} \tag{3}$$

Second, for the case in which the shifts for two isotopes of the same element ($Z_1 = Z_2 = Z$) are compared the ratio simplifies to

$$\frac{\rho_2}{\rho_1} = \frac{E_{\gamma_2}}{E_{\gamma_1}}\frac{F(Z,A_1)}{F(Z,A_2)}\frac{S_2}{S_1} \tag{4}$$

Both Eqs. (3) and (4) contain only well-known factors in addition to the experimental results. Thus, such ratios can be precisely determined.

For a case in which the shifts for two nuclei with isoelectronic chemical compounds are compared, expression (2) simplifies somewhat. Through detailed Hartree–Fock self-consistent-field (HF SCF) computations of electron densities for various ionic configurations, the ratio $\Delta\psi_{Z_1}{}^2/\Delta\psi_{Z_2}{}^2$ of the density differences between two configurations in each member of an isoelectronic pair is found to be nearly a constant. Furthermore, as will be shown later, this ratio is insensitive to the computational technique. Thus, denoting the ratio $\Delta\psi_{Z_1}{}^2/\Delta\psi_{Z_2}{}^2$ for any isoelectronic pair by $K(Z_1/Z_2)$, we have

$$\frac{\rho_2}{\rho_1} = \frac{E_{\gamma_2}}{E_\gamma} \frac{F(Z_1,A_1)}{F(Z_2,A_1)} K\left(\frac{Z_1}{Z_2}\right) \frac{S_2}{S_1} \tag{5}$$

which implies that the relation between shifts measured for two nuclei should be linear when proper isoelectronic pairs are selected. For example, typical isoelectronic pairs for various elements considered are: K_2SnF_6 *vs.* $KSbF_6$, K_2TeO_4 *vs.* KIO_4, $(IO_4)^-$ *vs.* XeO_4, and I^- *vs.* Xe°. In the actual use of this concept there is a shortage of cases, and in our anxiety to find additional pairs we include some "ionic" pairs for which it is evident from the chemical formula that the coordinations for the two cations differ. Examples of these less-certain pairings are: Sb_2O_3 *vs.* TeO_2, SnF_2 *vs.* SbF_3, and ICl *vs.* $XeCl_2$.

HF COMPUTATIONS

As mentioned earlier, two programs were used in computing *s*-electron density for various ionic configurations of each of the elements. These computations being nonrelativistic, these results have yet to be corrected by S′[6]. The justification for correcting

Table V. Values of Electron Density at the Nucleus for Some Electronic Configurations of Iodine*

Atomic configur-ation	$\psi^2(0)/a_0{}^3 - C$ Wilson–Froese program	$\psi^2(0)/a_0{}^3 - C$ Herman–Skillman program	$K(Z_2/Z_1) = \Delta\psi_{Z_2}{}^2/\Delta\psi_{Z_1}{}^2$ Wilson–Froese program	$K(Z_2/Z_1) = \Delta\psi_{Z_2}{}^2/\Delta\psi_{Z_1}{}^2$ Herman–Skillman program
$5s^2\,5p^5$	43.85	50.23	1.213	1.188
$5s^2\,5p^4$	47.31	53.95	1.214	1.191
$5s^2\,5p^3$	51.44	58.33	1.212	1.191
$5s^2\,5p^2$	56.01	63.33	1.208	1.189
$5s^2\,5p^1$	61.16	68.79	1.200	1.185
$5s^2\,5p^0$	66.67	74.32	1.195	1.183

*These values were computed with both the Wilson–Froese and the Herman–Skillman programs. The contribution C from closed inner shells is $C = 105,$ 401.8 and 105, 384.70, respectively, as computed from these two programs. The value of the ratio $\Delta\psi_{Z_2}{}^2/\Delta\psi_{Z_1}{}^2$ for xenon (Z_2) and iodine (Z_1) ions in the same configuration was obtained by each of the two computational programs.

by this simple technique, instead of doing detailed relativistic calculations, has been discussed at some length in an earlier work [2].

As it is impractical to repeat these computations in detail, a few of the many configurations computed for iodine are given as illustrations in Table V. For both xenon (Z_2) and iodine (Z_1) in a listed configuration Table V also gives the ratio $\Delta\psi_{Z_1}{}^2/\Delta\psi_{Z_2}{}^2$ as obtained with each of the programs. The closeness of average values for different configurations is remarkable and is true of all elements (Table VI) considered here. The averages have been obtained for 12–18 configurations.

Table VI. Average Ratio $\Delta\psi_{z_2}{}^2/\Delta\psi_{z_1}{}^2$ for Various Elements, As Obtained Through the Two Computational Programs

	$\Delta\psi_{Z_2}{}^2/\Delta\psi_{Z_1}{}^2$	
Element Z_2/Element Z_1	Wilson–Froese program	Herman–Skillman program
Sb/Sn	1.330	1.278
Te/Sb	1.237	1.238
I/Te	1.198	1.209
Xe/I	1.197	1.188

ρ FROM ISOELECTRONIC PAIRS

Figures 1–3 are plots of isomer shifts for isoelectronic compounds of elements differing in Z by unity, from Sb to Xe (the plot for Sn *vs.*

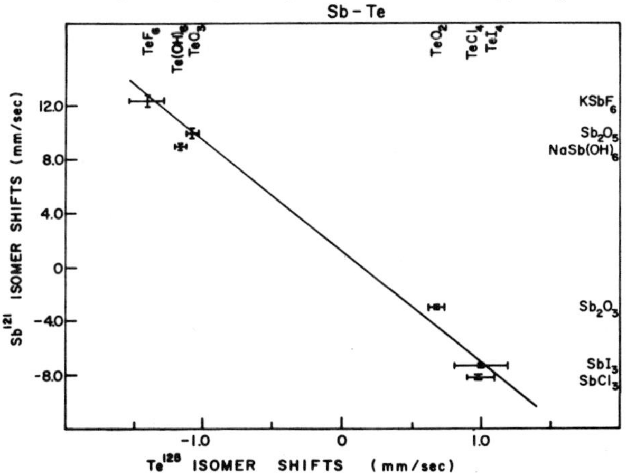

Fig. 1. Plot of isomer shifts of isoelectronic compounds of Sb^{121} and Te^{125}.

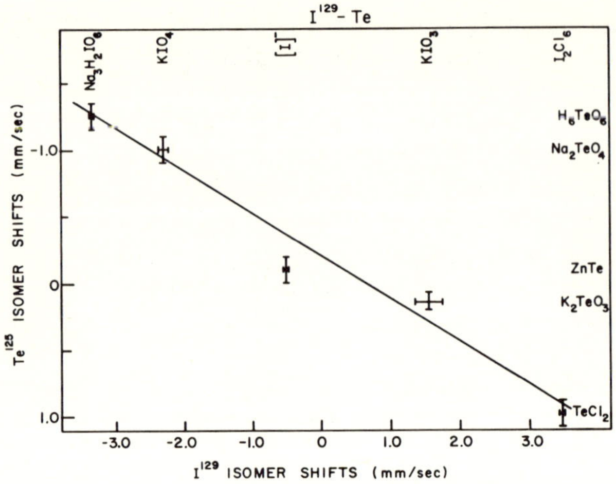

Fig. 2. Plot of isomer shifts of isoelectronic compounds of Te^{125} and I^{129}.

Sb is given in ref. [2]). In spite of the small number of isoelectronic pairs, the slope of the line is moderately well determined. The ratios of the values of ρ, obtained from the slopes of various plots with the aid of the Wilson–Froese values for $K(Z_1/Z_2)$ and relation (5), are given in Table VII. The estimated error in these ratios is about 15%.

For the sake of completeness we have also given the ratio of the values of ρ for I^{127} and I^{129}, for which Eq. (4) is applicable. The ratio is obtained by least-squares fitting the reported isomer shifts (listed in Table III). The value $\rho(I^{127})/\rho(I^{129}) = -0.78$ is in exact agreement with

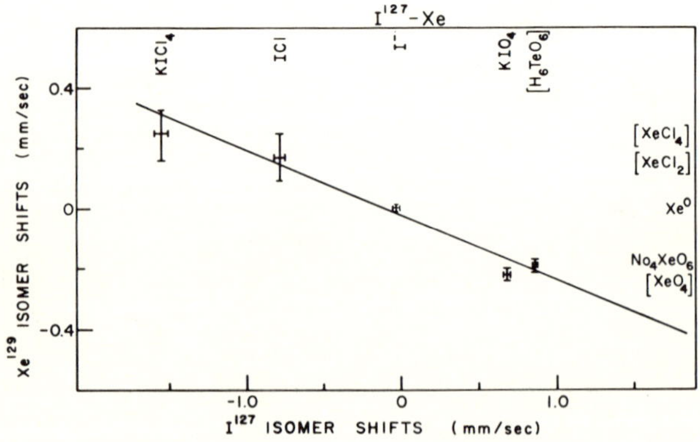

Fig. 3. Plot of isomer shifts of isoelectronic compounds of I^{127} and Xe^{129}.

Table VII. Ratios of the Values of ρ for Various Pairs of Nuclei

$[Z_1]/[Z_2]$	$\rho(1)/\rho(2)$
Sn^{119}/Sb^{121}	-0.16
Sb^{121}/Te^{125}	-7.6
Te^{125}/I^{129}	$+0.31$
I^{127}/I^{129}	-0.78
I^{127}/Xe^{129}	-8.7

Table VIII. Values of ρ for Various Nuclei

Nucleus	Values of $\rho \times 10^4$: Earlier work*	From comparison of isoelectronic structures	From density calibration
Sn^{119}	2.2, 2.4, 6.6, 1.84 [8]	—	2.4
Sb^{121}	-17.0, -36.8 [9]	-14.6	-17.0
Te^{125}	1.4 [6], 4.0, 0.48 [11], 7.2 [9]	1.9	—
I^{127}	-0.80, -0.56 [11]	-4.8	-5.6
I^{129}	1.0, 0.6	6.2	6.6
Xe^{129}	—	0.55	0.66

*References to earlier work can be found in ref. [24]. The references following a value refer to more recent work.

the result obtained by Reddy *et al.* [17]. A value of -0.65 was obtained by Jung and Triftshauser [11].

Taking the value of ρ for Sn^{119} to be 2.4×10^{-4}, as deduced by comparison of isomer shifts with calculated charge densities [2], we can compute the values of ρ for Sb^{121}, Te^{125}, I^{127}, I^{129}, and Xe^{129}. These values are listed in Table VIII, along with previously reported values by other authors [24].

ρ FROM DENSITY COMPUTATIONS

It was realized earlier that some insight into the electron distribution of an ion in a solid can be obtained from electric field gradients, which can be measured by NGR or NQR techniques, as well as from NMR chemical-shift measurements. Lees and Flinn [7, 25] related quadrupole splitting and the isomer shift in tin to assign specific isomer shifts to certain electronic configurations. In our earlier work [2] the electronic densities computed by use of HF SCF atomic wave functions for various configurations were used to calibrate these shifts, from which we obtained the value $\rho = 2.4 \times 10^{-4}$.

DeWaard *et al.* [19] have obtained the change in isomer shift per p hole for I^{129} by relating their isomer-shift values for alkali iodides to NMR chemical-shift studies. Perlow [14] has similarly obtained the change in isomer shift per p hole for I^{127} and Xe^{129} by using the data on the quadrupole interaction in various compounds of iodine and xenon. Having obtained the shift per removal of an s or a p valence electron, and, in addition, having calculated the change in electron density at the nucleus per s or p electron, one can deduce the value of ρ.

In their work on I^{129} DeWaard *et al.* [19] use the value of charge density obtained from the Fermi–Segré relation* to obtain $\rho = 0.6 \times 10^{-4}$. This is noticeably smaller than the value we obtain from the comparison of isoelectronic compounds. Our detailed computation of electron density indicates that the Fermi–Segré relation as used in this case gives too large a value for the density, and this in turn gives too small a value for ρ. Using our calculations and Hafemeister's expression relating the number of p holes and the change in isomer shift, we have recomputed the value of ρ. For configurations close to alkali iodides (namely, $5s^2\,5p^y$, $5 \leqslant y \leqslant 6$), the shift per p hole is 1.15 mm/sec. From this we obtain $\rho = 6.7 \times 10^{-4}$ for I^{129}.

Through his detailed analysis of isomeric shifts and quadrupole couplings for I^{127} Perlow [14] deduces values for the change in isomer shift per removal of an electron. His results are -0.56 mm/sec per $5p$ hole and $+3.40$ mm/sec per $5s$ electron.† Using our density computations independently for the two situations (either a $5p$ or a $5s$ hole), we obtain $\rho = -5.9 \times 10^{-4}$ and -5.1×10^{-4}, respectively. The internal consistency is impressive, and these values are in close agreement with our value obtained from isoelectronic pairs. However, it differs from the value of Jung and Triftshauser [11]. The reason for this discrepancy is again their use of the Fermi–Segré relation to compute the densities.

For Xe^{129} Perlow [14] finds that the change in isomer shift per p hole is $+0.13$ mm/sec. Using our density computations, we obtain $\rho = 0.66 \times 10^{-4}$ for this nucleus. Again this is in agreement with that obtained by comparing isoelectronic compounds.

*For detailed discussion see Kopfermann [26].

†For I^{127} the shift per $5p$ or $5s$ electron has also been reported by other authors. M. Kaplan (private communication) obtains a value of -0.50 mm/sec/$5p$ hole. The values reported by Pasternak [20] for I^{129} when converted to I^{127} are -0.51 mm/sec/$5p$ hole and 3.1 mm/sec/$5s$ electron. Recently Bukhspan *et al.* [15] report -0.56 mm/sec/$5p$ hole from their measurements on I^{127} in frozen nonpolar solutions.

For Te^{125}, on the other hand, it is hard to obtain the information on the shift per electron for two reasons. Not only is tellurium chemistry complex, but also the accurate measurement of quadrupole interactions is hard, since the natural width of the lines is usually comparable to or greater than the quadrupole splitting [11]. Also, it is unfortunate that none of the tellurium isotopes are suitable for NQR measurements.

Regarding Sb^{121}, Shpinel *et al.* [27] have recently carried out an analysis comparing shifts for halide complexes of tin and antimony much as in ref. [2]. They arrive at a value $\rho(\mathrm{Sb})/\rho(\mathrm{Sn}) = -6$, which is close to the earlier value -6.1. However, in deducing the absolute value of ρ for Sb^{121}, they adopted the value $\rho(\mathrm{Sn}^{119}) = 6.6 \times 10^{-4}$ obtained from conversion-electron measurements [28]. In view of the rather large uncertainties associated with this beautiful (but difficult) experiment, we prefer to use the "classical" [2] value 2.4×10^{-4}.

Once the value of ρ has been obtained for various nuclei, it is useful to obtain a correlation between electron density and the isomer shifts. From the value of electron density calculated for a variety of ionic configurations it is possible to draw such a correlation plot once we decide on the electron distribution of the ion in any one of the compounds. Earlier we had assumed that α-tin and InSb both give rise to an sp^3 configuration. We shall assume the electron distribution on I^- and Xe° to be $5s^25p^6$. Figures 4 and 5 are such correlation plots for

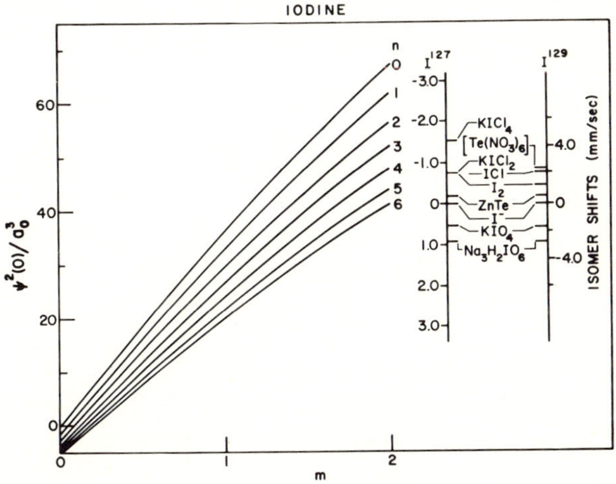

Fig. 4. Correlation between electron density $\psi^2(0)/a_0{}^3$ at the nucleus and the isomer shift for I^{127} and I^{129} compounds as a function of the number m of $5s$ electrons. The parameter n is the number of $5p$ electrons.

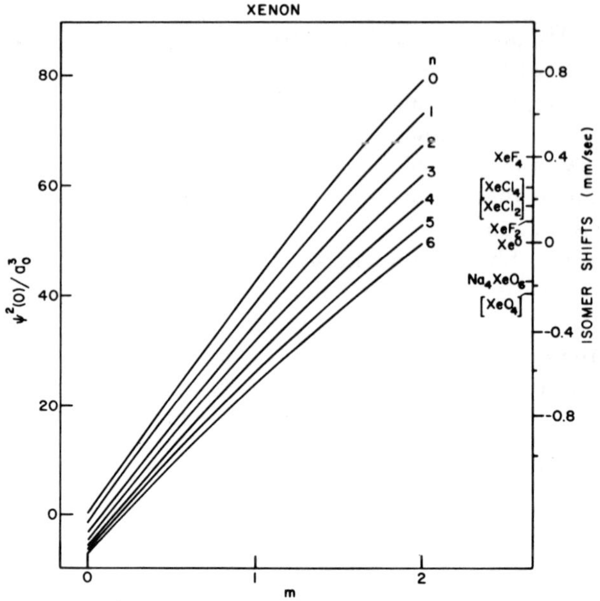

Fig. 5. Correlation between electron density at the nucleus and isomer shift, as in Fig. 4, for the xenon compounds.

iodine and xenon compounds, where the density values used are from the Wilson–Froese program. It is worth noting that the configurations predicted for various compounds on the basis of these plots closely agree with those deduced by Perlow [14] through the interpretation of quadrupole interaction. Similar plots for Sn^{119} and Sb^{121} compounds have been given in the earlier paper [2].

For Te^{125}, however, we have obtained the correlation (Fig. 6) between shifts and the density by assuming the value $\rho = 1.9 \times 10^{-4}$ deduced from the comparison between isoelectronic compounds of Sn and Te, and assuming the tellurium ion to have a $5s^2 5p^6$ configuration in ZnTe. Such calibration gives a shift of 0.30 mm/sec per p hole and -1.8 mm/sec per s electron.

CONCLUSIONS

The value of ρ obtained from isomer-shift measurements depends directly on the knowledge of the s-electron density at the nucleus. It seems that more-reliable values for the densities can be obtained from HF SCF atomic calculations than from the use of the Fermi–Segré relation. The large difference between the values of ρ for iodine

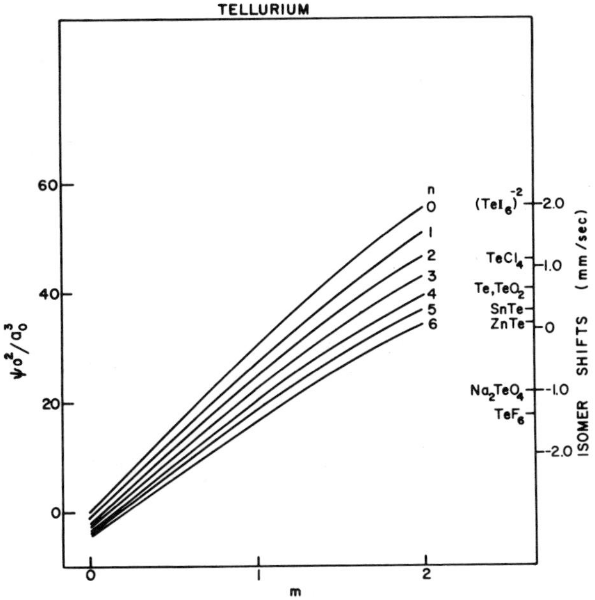

Fig. 6. Correlation between electron density at the nucleus and isomer shift, as in Fig. 4, for the tellurium compounds.

and tellurium isotopes obtained in this work and those from the earlier works is a clear reflection of this.

The value of ρ obtained for Sn^{119} by use of Waber's relativistic wave functions as reported by Lees and Flinn [7] is within the error limits of our value. This gives considerable confidence in correcting nonrelativistic densities to relativistic ones by applying the factor $S'(Z)$—at least for the isotopes considered in this paper. It is pertinent to mention that when the relativistic value of $\psi_{ns}{}^2$ for Sn is divided by the corresponding Herman–Skillman value, the ratio is 2.17 for $1 \leqslant n \leqslant 4$. This is to be compared with the value $S'(50) = 2.31$ reported by Shirley [6]. However, the same ratio for the $5s^2$ density becomes 2.59. Such discrepancies probably reflect the deficiencies at the current stage of development of these calculations.

A remark must be made on more realistic calculations of electron densities using molecular orbitals in comparison with our own HF SCF atomic calculations. For the tin compounds Unland and Letcher [29] have carried out an elaborate density computation in which they specify molecular wave functions. This approach was found to be highly sensitive to wave-function parameters, but it yielded a value for ρ close to the "classical" value [2] for Sn^{119}. In spite of the fact that such a procedure yielded a value close to that deduced from simple atomic functions, it is worthwhile to make similar attempts on other elements.

The technique of pairing isoelectronic compounds seems fruitful in view of the fact that the final results are fairly independent of the uncertainties in density computations. The values of ρ obtained in such a fashion have been close to those obtained by use of density calibration. Even though this is a heartening result in view of the uncertainty involved either in isoelectronic pairing of two compounds or the absolute values of computed densities, the error in the values of ρ given in Table VIII is estimated to be about 20%.

In the past, isomer shifts have been expressed in terms of number of p holes (N_p) and s holes (N_s) in the form $S = S_0 + aN_s + bN_p$. Such linear dependences are superseded by the graphs shown in Figs. 4–6 (and by similar graphs for Sn and Sb given in ref. [2]). It is clear that b would be approximately halved if evaluated near $N_s = 1$ as compared with $N_s = 2$. Similarly, a changes, although less drastically, as a function of N_p. On the other hand, if one stays near a given configuration like $5s^2 5p^6$, then such coefficients perform a useful role.

For I^{127} we obtain $S = -0.56N_p + 3.3N_s$. For Xe^{129} compounds we obtain $S = 0.11N_p - 0.66N_s$. A similar result for Te^{125} is $S = 0.1 + 0.25N_p - 1.5N_s$. More-complex relations depending on higher powers of N_s and N_p have been given for Sn^{119} by Lees and Flinn [7].

We shall not say much about the comparison of our values of ρ with those obtained from nuclear theory. Uher and Sorensen [30] have made an attempt to calculate this delicate property in a general way. Using a "pairing-plus-quadrupole" model, they have obtained results which are nearly like ours for Sn^{119}, but are much smaller for $I^{127,129}$. It is hoped that future work will clarify the significance of our values for ρ.

ACKNOWLEDGMENTS

We are greatly indebted to Dr. Michael Wilson for acquainting us with SCF atomic calculations. Bruce Zabransky and Joe Kulaga assisted greatly in work in the laboratory.

REFERENCES

1. P. Kienle, in: *Hyperfine Structure and Nuclear Radiation*, E. Mathias and D. A. Shirley, eds. (North-Holland Publishing Co., Amsterdam 1968), p. 27.

2. S. L. Ruby, G. M. Kalvius, G. B. Beard, and R. E. Snyder, *Phys. Rev.* **159**:239 (1967).
3. F. Herman and S. Skillman, *Atomic Structure Calculations* (Prentice-Hall, Englewood Cliffs, New Jersey, 1963).
4. D. Liberman, J. T. Waber, and D. T. Cromer, *Phys. Rev.* **137**:A27 (1965).
5. C. Froese, *J. Chem. Phys.* **45**:1417 (1966).
6. D. A. Shirley, *Rev. Mod. Phys.* **36**:339 (1964).
7. J. K. Lees and P. A. Flinn, *J. Chem. Phys.* **48**:882 (1968).
8. S. L. Ruby and G. K. Shenoy, *Phys. Rev.* October (1969).
9. V. S. Shpinel, V. A. Bryukhanov, V. Kothekar, B. Z. Iofa, and S. I. Semenov, in: *The Mössbauer Effect* (Symposium of the Faraday Society, No. **1** 1967) (Plenum Press, New York, 1968), p. 69.
10. C. E. Violet and R. Booth, *Phys. Rev.* **144**:225 (1966).
11. P. Jung and W. Triftshauser, *Phys. Rev.* **175**:512 (1968).
12. A. B. Buyrn and L. Grodzins, *Bull. Am. Phys. Soc.* **8**:43 (1963).
13. J. F. Ullrich, Thesis, University of Michigan, July 1967.
14. G. J. Perlow, in: *Chemical Applications of Mössbauer Spectroscopy*, V. I. Goldanskii and R. H. Herber, eds. (Academic Press, New York, 1968), p. 377.
15. S. Bukhspan, C. Goldstein, and T. Sonnino, *J. Chem. Phys.* (to be published).
16. G. J. Perlow and S. L. Ruby, *Physics Letters* **13**:198 (1964).
17. K. R. Reddy, F. DeBarros, and S. DeBenedetti, *Physics Letters* **20**:297 (1966).
18. D. W. Hafemeister, G. DePasquali, and H. DeWaard, *Phys. Rev.* **135**:B1089 (1964).
19. H. DeWaard, G. DePasquali, and D. W Hafemeister, *Physics Letters* **5**:217 (1963).
20. M. Pasternak, in: *The Mössbauer Effect* (Symposium of the Faraday Society, No. 1, 1967) (Plenum Press, New York, 1968), p. 119.
21. M. Pasternak, A. Simopoules, and Y. Hazony, *Phys. Rev.* **140**:A1892 (1965).
22. M. Pasternak and T. Sonnino, International Conference on Hyperfine Nuclear Spectroscopy, Wellington, New Zealand, 1966.
23. M. Pasternak and S. Bukhspan, *Phys. Rev.* **163**:297 (1967).
24. P. Kienle, G. M. Kalvius, and S. L. Ruby, in: *Hyperfine Structure and Nuclear Radiation*, E. Mathias and D. A. Shirley, eds. (North-Holland Publishing Co., Amsterdam, 1968), p. 971.
25. J. K. Lees and P. A. Flinn, *Physics Letters* **19**:186 (1965).
26. H. Kopfermann, *Nuclear Moments* (Academic Press, New York, 1958), p. 144.
27. V. S. Shpinel, V. A. Bryukhanov, V. Kothekar, and B. Z. Iofa, *Soviet Phys.—JETP* **26**:16 (1968).
28. J. P. Bocquet, Y. Y. Chu, O. C. Kistner, M. L. Perlman, and G. T. Emery, *Phys. Rev. Letters* **17**:809 (1966).
29. M. L. Unland and J. H. Letcher, *J. Chem. Phys.* **49**:2706 (1968).
30. R. A. Uher and R. A. Sorensen, *Nucl. Phys.* **86**:1 (1966).

Mössbauer Studies of Vitamin B_{12} and Some Related Cobalamins

R. T. Mullen

Radioactivity Laboratory
Merck Sharp & Dohme Research Laboratories
Division of Merck & Co., Inc.
Rahway, New Jersey

The Mössbauer spectra of vitamin B_{12}-Co^{57} (*cyanocobalamin*-Co^{57}), *dicyanocobalamin, and several aquocobalamin derivatives have been investigated. Sources were prepared as thin polycrystalline solids or frozen aqueous solutions and run, mainly at* 77°K, *against a* $Na_4Fe(CN)_6 \cdot 10H_2O$ *absorber. The spectra of these cobalamins consist in general of a fairly strong, broad* (~1 mm/sec *half-width*) *doublet with evidence in many, but not all, cases of a minor contribution from a second doublet. Quadrupole splittings vary from* 1.8 to 3.1 mm/sec; *isomer shifts* (vs. *SNP*) *are largely in the* +0.25–0.55 mm/sec *range. Chemical changes in dicyanocobalamin solutions are observable by Mössbauer spectroscopy, although undetected by ultraviolet spectroscopy. The similarity of the Mössbauer spectrum of vitamin* B_{12} *to those of certain other biologically active compounds, particularly cytochrome c, is discussed, as is the general problem of interpreting "source" experiments.*

INTRODUCTION

Rather early in the history of the use of the Mössbauer effect in chemical studies Gonser *et al.* [1] and other workers [2] reported on the Mössbauer spectroscopy of hemoglobin and related porphyrins. More recently, Cooke and Debrunner [3] described similar investigations of cytochrome c. The Mössbauer spectra of some cobalamins and other "source" compounds have been studied by Nath *et al.* [4], who have as their main interest the determination of the Mössbauer consequences of the Auger electron processes in the host materials.

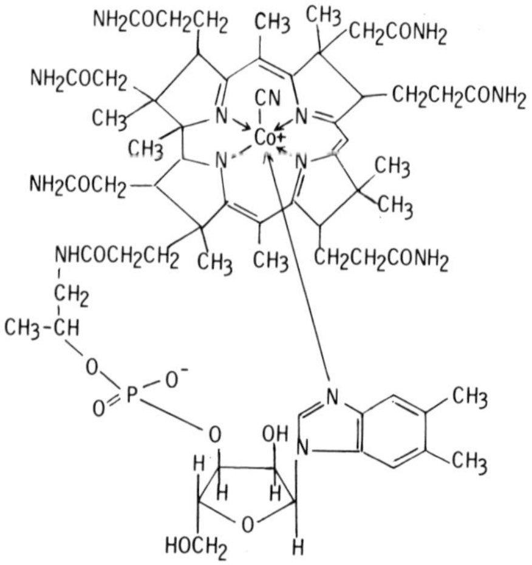

Fig. 1. Structure of vitamin B_{12}.

Thus, the application of Mössbauer spectroscopy to the study of compounds of biological interest is not novel, but, limited for the most part to iron-containing compounds, neither has it become widespread.

The work described here was undertaken primarily as an experiment in methodology—to determine from our own experience the role Mössbauer spectroscopy might play in a laboratory primarily concerned with organic molecules of biological significance. The vitamin B_{12} (cyanocobalamin) family was chosen for study because of its biological importance and significance to our laboratory. That we were not deterred from Mössbauer studies of a cobalt compound of interest to us because of the possible complications due to the Auger events may be of more general significance and will be discussed later.

The structure of vitamin B_{12} is shown in Fig. 1. The corrin ring of the cobalamins is closely related to porphine, differing mainly in the absence of a bridging carbon atom between the *A* and *D* rings. The cobalt ion is diamagnetic [low-spin Co(III)] and very firmly bound [5]. The other cobalamins discussed in this work differ from cyanocobalamin primarily with respect to one or both of the axial cobalt ligands.

EXPERIMENTAL

A constant-acceleration spectrometer employing a 400-channel analyzer (RIDL 34–12B) was used. The technique of Kankeleit [6] was followed for the generation of the reference wave and the method of Cohen *et al.* [7] used for the remainder of the feedback loop. The drive–sensor unit was constructed from two loudspeakers (University C-8HL) mounted vertically back-to-back. The drive rod terminated in a ring through which the detector (a Reuter-Stokes Xe–CH_4 proportional counter) was positioned. A 1 mg/cm^2 $Na_4Fe[CN]_6 \cdot 10H_2O$ absorber was mounted on the front face of the ring. Cobalamin sources were put in copper, magnesium, or Kel-F holders fastened to the bottom of the cryostat tailstock. The source–absorber distance was 4.5 in. The maximum deviation of the gamma-ray beam from the vertical was 5.5°. Calibration was made by running a 2-mCi Co in Pd source (NENC) against a natural iron foil. Toward the end of this work a single-crystal NBS sodium nitroprusside absorber was also used for calibration.

Sources consisting of several hundred microcuries of Co^{57} in 0.1–0.8 mg solid were prepared from cyano- or hydroxocobalamin-Co^{57} (Merck & Co., Inc.) and run as the polycrystalline solid or in frozen solution ($\sim$ 70 μl volume). Purity was monitored by thin-layer chromatography and paper chromatography at various stages of preparation and again after individual runs.

RESULTS

Isomer shifts and quadrupole splittings are listed in Table I. A scatter plot of our results together with those of Nath *et al.* [4] for the cobalamins and Cooke and Debrunner [3] for oxidized cytochrome c is shown in Fig. 2. It will be noted that the results for oxidized cytochrome c, both in the solid and in the frozen aqueous solution, are quite similar to the respective cyanocobalamin data. Although not shown in Fig. 2, a number of low-spin hemoglobin derivatives also fall in the same general area. Slight deviations from theoretical symmetry were noted in some of the spectra. Semiquantitative analysis made with a DuPont Curve Resolver indicated that these deviations could be explained by the contribution of a second doublet of an intensity 5–15 % of the major doublet. Digital-computer analysis now in progress will permit a more precise evaluation of this possibility. Preliminary results show good agreement between the relative intensity of the minor doublet needed to reproduce an experimental spectrum and the degree of heterogeneity induced in some samples by handling and radiation damage as seen by thin-layer chromatography.

Table I. Cobalamin Mössbauer Data*

Sample	Temp., °K	Isomer shift, mm/sec	Quadrupole splitting, mm/sec	No. runs
Cyanocobalamin	77	0.39 ± .01	2.07 ± .01	3
Cyanocobalamin	195	0.35	2.02	1
Cyanocobalamin	295	0.26	1.94	1
Cyanocobalamin aqueous solution	77	0.36 ± .02	2.13 ± .08	7
Dicyanocobalamin aqueous solution	77	0.30 ± .02	1.86 ± .02	2
Dicyanocobalamin aqueous solution	195	0.31	1.78	1
Dicyanocobalamin aqueous solution (thermally exposed)	77	0.56 ± .04	2.58 ± .04	2
Hydroxocobalamin	77	0.37	3.06	1
Acetatocobalamin	77	0.47	2.50	1
Acetatocobalamin methanol solution	77	0.50	2.47	1
Chlorocobalamin	77	0.45	3.07	1
Reduced cyanocobalamin on platinum	77	0.55	2.72	1

*Isomer shifts are with respect to sodium nitroprusside, calculated by addition of 0.20 mm/sec to the isomer shifts determined experimentally *vs.* sodium ferrocyanide decahydrate. All samples were run as the solid unless otherwise indicated. The uncertainties listed are the observed average deviations. The dicyanocobalamin samples were frozen promptly after preparation except where noted.

Cyanocobalamin

A spectrum of solid cyanocobalamin is shown in Fig. 3. The quadrupole splitting is large; its magnitude and that of the isomer shift are consistent with a low-spin Fe(III) assignment. The temperature dependence of the quadrupole splitting exhibits a slight negative slope. This too is consistent with low-spin Fe(III) based on the arguments given by Cooke and Debrunner [3] for that assignment in oxidized cytochrome c. Thus, the oxidation number and low spin state of cobalt in vitamin B_{12} appear to be maintained in its daughter iron compound.

Dicyanocobalamin

Once the Mössbauer parameters of cyanocobalamin were established it was thought of interest to apply this technique to an examination of a closely related derivative, dicyanocobalamin. This compound is formed readily upon treatment of cyanocobalamin with aqueous cyanide solution. A prompt color change from red to purple

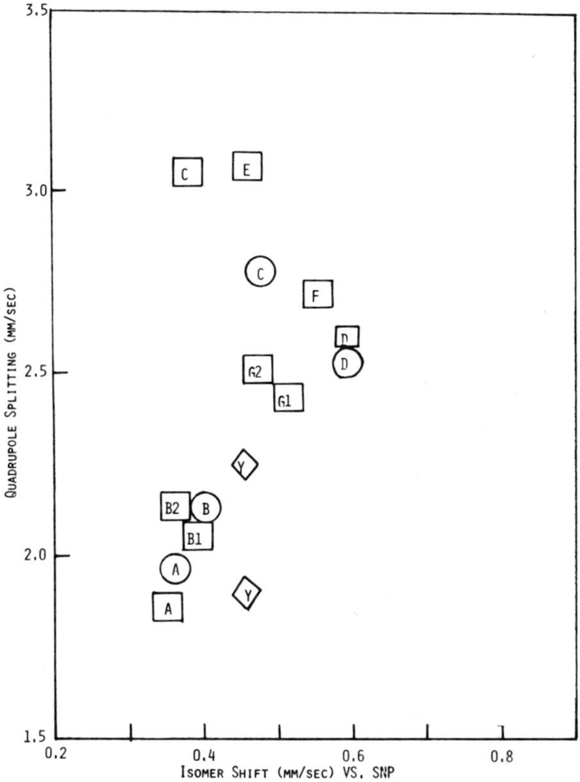

Fig. 2. Quadrupole splitting *vs.* isomer shift for cobalamins and oxidized cytochrome c. Squares: the present work at 77°K; *A*: dicyanocobalamin aqueous solution; *B*1: solid cyanocobalamin; *B*2: cyanocobalamin aqueous solution; *C*: hydroxocobalamin; *D*: thermally exposed dicyanocobalamin; *E*: solid chlorocobalamin; *F*: reduced cyanocobalamin in platinum; *G*1: acetocobalamin methanol solution; *G*2: solid acetatocobalamin.
Circles: Nath *et al.* [4], solids at 77°K (+0.20 mm/sec added to reported isomer shifts *vs.* sodium ferrocyanide); *A*: dicyanocobalamin; *B*: cyanocobalamin; *C*: aquocobalamin; *D*: ammonocobalamin. Diamonds: Cooke and Debrunner [3] at 100°K; *X*: oxidized cytochrome c solid; *Y*: oxidized cytochrome c aqueous solution.

occurs. An interpretation of the ultraviolet spectrum of dicyanocobalamin by Beaven *et al.* [8] ascribes the formation of the dicyanocobalamin derivative to the displacement of the benzaminazole coordinate linkage of vitamin B_{12} by a second cyanide ion.

On running these spectra, it was found that two extreme cases resulted, depending on the thermal history of the sample. If the dicyanocobalamin solution is frozen promptly (5–10 min) after

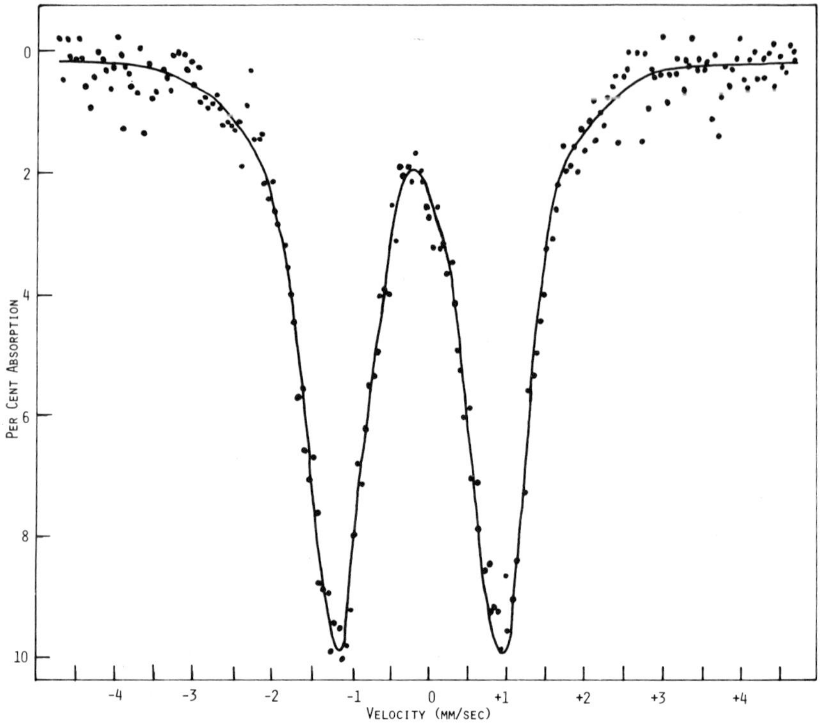

Fig. 3. Mössbauer spectrum of solid vitamin B_{12} *vs.* sodium ferrocyanide decahydrate at 77°K.

preparation, a spectrum is observed that is very similar to that of vitamin B_{12} and in good agreement with that obtained by Nath *et al.* [4] for solid dicyanocobalamin (see Fig. 2). The isomer shift and quadrupole splitting are slightly smaller than those of vitamin B_{12} in frozen aqueous solutions—changes in the direction expected from the greater symmetry and addition of a negative charge upon substitution of cyanide anion for the neutral benzaminazole coordinate bond.

However, if the dicyanocobalamin solution (0.16*M* in NaCN or KCN) is exposed to room temperature for several hours, a spectrum with quite different Mössbauer parameters is observed as shown in Fig. 4. In this case the isomer shift and quadrupole splitting are both substantially larger than in vitamin B_{12}, indicating a lower electron

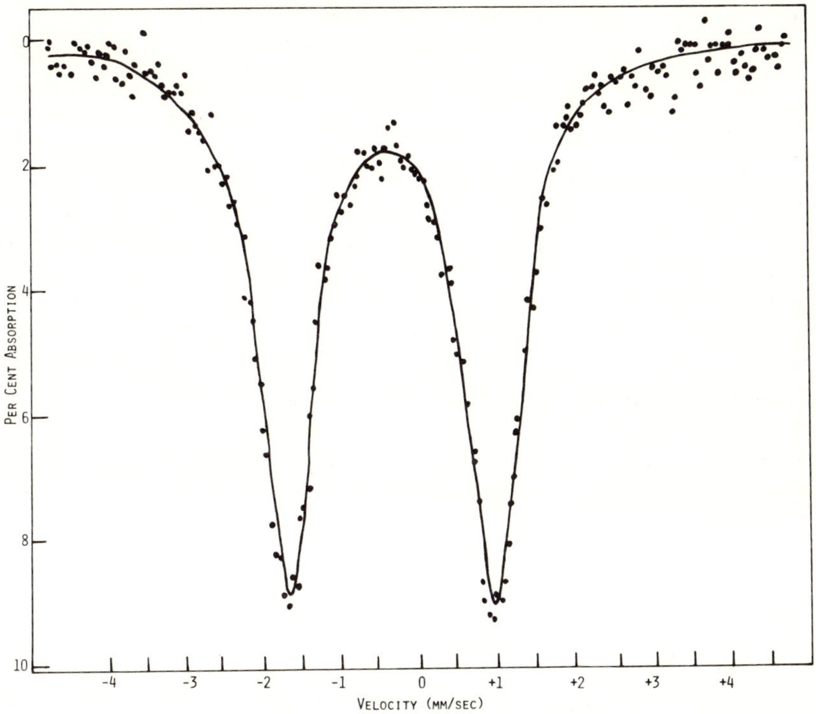

Fig. 4. Mössbauer spectrum of thermally exposed dicyanocobalamin aqueous solution *vs.* sodium ferrocyanide decahydrate at 77°K.

density and a higher degree of electric field asymmetry at the metal nucleus in the thermally exposed dicyanocobalamin samples. The ability to recover cyanocobalamin from such a sample upon acidification and purification is evidence that we are at least in part observing a slow transition within the "dicyanocobalamin" system rather than an irreversible gross degradation. These results may be related to earlier observations by Conn *et al.* [9] in these laboratories that the equilibrium between cyanide ion and cyano-dicyanocobalamin is more complex than originally believed.

We were unable to observe this phenomenon by ultraviolet spectroscopy in a series of controlled exposures of dicyanocobalamin solutions to room temperature in various Mössbauer sample holders. This could be due to the necessity to dilute the samples prior to ultraviolet analysis. Whatever the reason, the significance for Mössbauer methodology is the ease with which this effect could be detected by Mössbauer spectroscopy.

Aquocobalamins

Replacement of the cyanide ligand in cyanocobalamin by a hydroxyl group gives rise to hydroxocobalamin. This compound may be regarded as the parent of a series of closely related aquocobalamin derivatives which are obtained upon treatment of hydroxocobalamin with appropriate acids. The spectrum of acetatocobalamin in frozen methanol at liquid-nitrogen temperature is shown in Fig. 5. It will be seen from Table I that significant differences exist in isomer shift and quadrupole splitting between acetatocobalamin and chlorocobalamin. Assuming that the iron analogs reflect the structure of the parent cobalt compounds, these differences are surprising if the acetate and chloride moieties are merely counter ions, but could be more readily explained if one or both of these groups are themselves directly attached to the cobalt atom. The spectrum of reduced cyanocobalamin

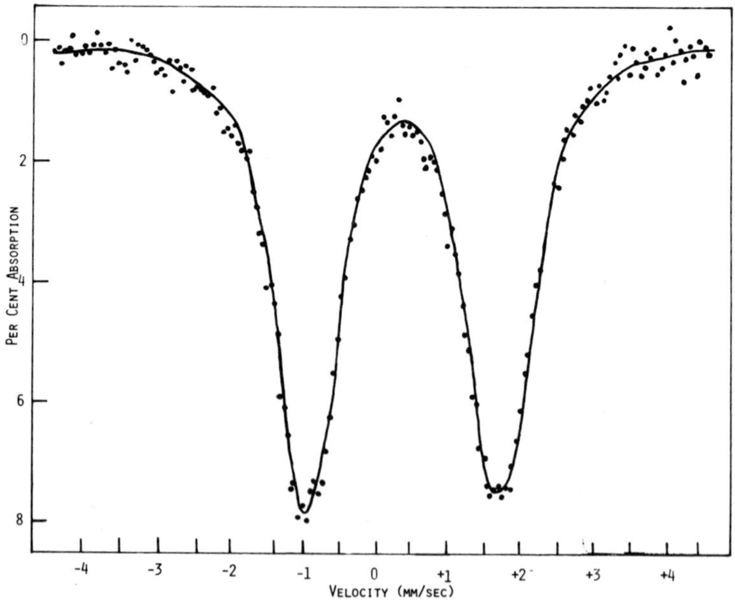

Fig. 5. Mössbauer spectrum of acetatocobalamin in methanol *vs.* sodium ferrocyanide decahydrate at 77°K.

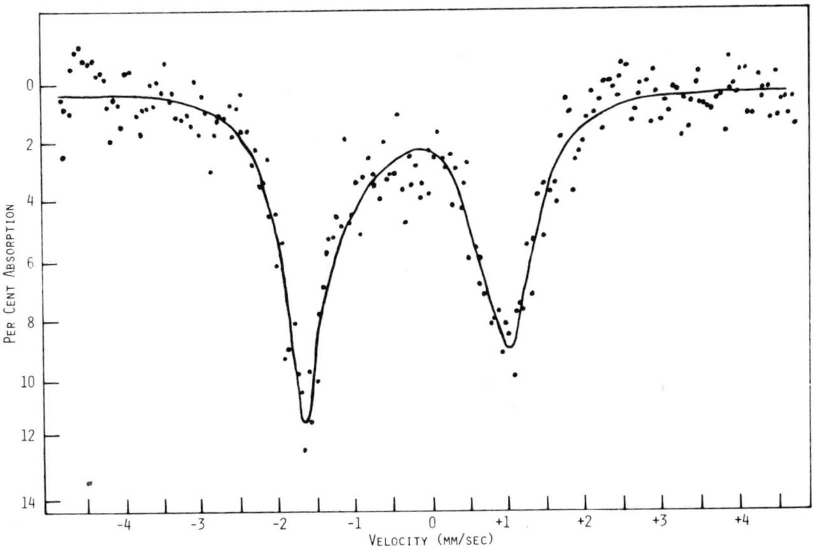

Fig. 6. Mössbauer spectrum of reduced cyanocobalamin on platinum catalyst *vs.* sodium ferrocyanide decahydrate at 77°K.

absorbed on platinum catalyst is shown in Fig. 6. Not itself an aquocobalamin, it was produced in an abortive attempt to synthesize hydroxocobalamin. Its spectrum is anomalous in the asymmetry of the doublet intensity and half-width. This may be ascribed to the highly anisotropic bonding of the molecule on the surface of the catalyst.

DISCUSSION

Both our results and those of Nath *et al.* [4] support the thesis that there is no "explosion" of the parent compound as a result of multiple Auger events in the cobalamins studied. The consistent heterogeneity expected from such a drastic occurrence is not observed. Further, the Mössbauer parameters of these cobalamins are similar to those reported for iron compounds of similar structure.

As Nath *et al.* [4] have pointed out, there is one major advantage to a Co^{57}-source experiment: when source compounds with high specific activities can be prepared, it is possible to minimize extraneous absorption of the Mössbauer radiation by nonresonant atoms. This can be especially valuable when applied to compounds of biological interest, as these materials are often of high molecular weight and contain the Mössbauer nuclide at high dilution.

The price one pays for this asset depends on whether one's main interest is in the daughter iron compound or the cobalt parent. If the former, then one is concerned not only with the possibility of major disruption of the molecule as a result of the nuclear transformation and its extranuclear aftermath, but also that the iron atom relaxes to the desired state. Should this be established—ideally, by running mirror Fe and Co spectra—one may well prefer the source approach.

In the present work, where interest lies chiefly in the Co^{57} source, a more-fundamental difficulty is encountered. As one is inevitably confined to observations of Fe^{57} daughter, one can relate the quantitative information extractible from a Mössbauer spectrum to the parent compound only in an indirect or inferential manner, and then with great caution. We note with interest the similarity of the Mössbauer spectra of vitamin B_{12} and cytochrome c in view of parallels in some of their biological functions, e.g., participation in electron-transport reactions. Indeed, Beaven *et al.* [8] suggested in 1950 that vitamin B_{12} might function in such reactions because of structural similarities it has to cytochrome c. In the light of current knowledge, there are difficulties quite apart from Mössbauer spectroscopy in any attempt to read significance into these parallels. However, this limitation of Mössbauer emission spectroscopy further complicates matters by requiring the implicit assumption that the iron analog of vitamin B_{12} would have a similar biological function to that of vitamin B_{12} itself.

Awareness of the potential difficulties of interpretation of certain Mössbauer experiments should not obscure the important role this technique, in common with many other kinds of spectroscopy, may play as a fingerprint method. Our experience with dicyanocobalamin indicates that Mössbauer emission spectroscopy can indeed function usefully in this manner. It is true that Mössbauer spectra are usually not so rich in information or structure as, for example, infrared or nuclear magnetic resonance spectra. This, of course, arises from the high selectivity of Mössbauer spectroscopy, and rather than being a liability, can be its greatest asset in biochemical applications. Certainly the tunnel vision inherent in this technique—its ability to focus on the Mössbauer nuclide to the relative exclusion of more remote and less interesting groups in a molecule—has been the key to its utility in studying macromolecules such as hemoglobin.

ACKNOWLEDGMENT

The author would like to gratefully acknowledge helpful conversations with Drs. Wertheim, Cohen, and Herber at the onset of this work. Acknowledgment is also due Mr. Michael Kniazuk, Mr. Andrew Reich, and Mr. Robert Jones for their many important contributions. The sustained interest and support of this research by Dr. Charles Rosenblum is most appreciated.

REFERENCES

1. U. Gonser, R. W. Grant, and J. Kregzde, *Science* **143**:680 (1964).
2. Y. S. Moshkovskii, "Applications of the Mössbauer Effect in Biology," in: *Chemical Applications of Mössbauer Spectroscopy*, V. I. Goldanskii and R. M. Herber, eds. (Academic Press, New York, 1968), Chap. 10.
3. R. Cooke and P. Debrunner, *J. Chem. Phys.* **48**:4532 (1968).
4. A. Nath, M. Harpold, M. P. Klein, and W. Kündig, *Chem. Phys. Letters* **2**:471 (1968).
5. E. L. Smith, *Vitamin B_{12}* (John Wiley & Sons, New York, 1965), Chap. 4.
6. E. Kankeleit, *Rev. Sci. Instr.* **35**:194 (1964).
7. R. L. Cohen, P. G. McMullin, and G. K. Wertheim, *Rev. Sci. Instr.* **34**:671 (1963).
8. G. H. Beaven, E. R. Holiday, E. A. Johnson, B. Ellis, and V. Petrow, *J. Pharm. Pharmacol.* **2**:944 (1950).
9. J. B. Conn, S. L. Norman, and T. G. Wartman, *Science* **113**:658 (1951).

Applications

Polarization Effects in Mössbauer Absorption by Single Crystals

R. M. Housley

Science Center
North American Rockwell Corporation
Thousand Oaks, California

There are a number of reasons for analyzing Mössbauer intensity measurements in single crystal samples. By studying the total intensity as a function of direction in the crystal, one can hope to obtain the six parameters that describe the mean square displacement at a resonant site. By studying relative intensities in a nonmagnetic crystal, one can hope to obtain the three or four parameters describing the electric field gradient at a resonant site which are not determined by the line positions. In a magnetic crystal one can hope to find the internal field directions and hence to learn a great deal about the magnetic structure.

Since in practice it is rarely possible to work with single crystal absorbers thin enough that saturation effects can be neglected, it is necessary to have a method of analysis in which saturation is taken into account. In a single crystal the absorption cross section at any energy will, in general, be partially polarized. It is easy to see that this will affect the saturation of the absorption. For example, consider an unpolarized incident beam and a completely polarized absorber. No matter how thick the absorber is made, in this case only half the incident beam can be resonantly absorbed. Corresponding to each absorption resonance is a contribution to the real part of the index of refraction having the same polarization but a different energy dependence. These lead to birefringence effects, which, by altering the polarization of radiation propagating through the crystal, can in turn affect the absorption. Blume and Kistner have described an elegant method for treating both these effects by using a 2×2 *complex matrix to represent the index of refraction of the absorber. If the internal field parameters and the mean square displacements in an absorber are known, the transmitted intensities can be calculated in a straightforward manner.*

In the applications described above, it is necessary to go in the other direction and obtain the internal field parameters and mean square displacement values from the measured intensities. This is greatly simplified in several cases which will be described. The most important case is if measurements are made along certain high symmetry directions in otherwise arbitrarily complicated crystals. In these cases the incident gamma-ray beam may be decomposed into two components, each propagating through the crystal with its own complex index of refraction.

The deviation of the index of refraction from one may be regarded as due to the part of the incident radiation coherently scattered in the forward direction. Conditions for neglecting effects of coherent scattering in other directions are described.

Use of this analysis will be illustrated with several examples. These include determination of the mean square displacement tensor for Fe *in siderite, determination of the mean square displacement and refinement of the electric field gradient for* Fe *in sodium nitroprusside, and explanation of some apparent anomalies which have appeared in the literature.*

INTRODUCTION

Mössbauer absorption measurements on single-crystal absorbers have been used recently for at least three purposes: (1) to determine the orientation of the principal axes of the electric field gradient (efg) tensor at Fe sites in complicated crystals [1–5]; (2) in at least two cases to determine the tensor which describes the mean square displacement (msd) of Fe ions in a crystal [5, 6]; and (3) to obtain information about the magnetic structure of several compounds [7–9].

These measurements all have one factor in common. It is the intensities or relative intensities of the absorption lines rather than their positions which provide the useful information. Therefore, in order to interpret data for these purposes, it is necessary to have an accurate theory giving the absorption dips as a function of crystal thickness and the efg, internal field, and msd parameters. This is not trivial, because of the polarization dependence of the individual nuclear cross sections. Because of this, the depths [10] or areas [11] of absorption lines cannot be calculated by the simple formulas used for unpolarized lines, except in the limit of very thin crystals.

Other measurements in which intensities are frequently important and cannot be calculated without taking polarization effects into account are all measurements made in applied magnetic fields. For example, polarization effects must be taken into account in making a detailed interpretation of spectra obtained by applying a magnetic field to a thick, randomly oriented powder of a material exhibiting a quadrupole doublet [12].

Two fairly complete and elegant solutions have been given to the problem of calculating the intensities transmitted through an absorber of arbitrary thickness in the case where all the parameters describing the individual nuclear cross sections are known. The first treatment, by Blume and Kistner [13], introduces a complex matrix index of refraction to describe the material. Boundary conditions are not discussed, but they are probably only important near grazing incidence. Also, the treatment does not describe propagation in the vicinity of a Bragg angle, but otherwise appears to be quite general.

We have made a generalization of the Blume and Kistner treatment which applies to crystals with more than one resonant site per unit cell [14]. This was done by introducing an average scattering amplitude per unit cell obtained by summing the properly weighted scattering amplitudes from all sites, all scattering amplitude matrices being expressed in terms of the same basis polarizations. A further generalization to randomly oriented fine powders is possible by merely making a properly weighted average over all orientations of the grains. This might be useful in analyzing spectra in cases where magnetic fields have been applied to powders.

We found that the formulas relating absorption as a function of thickness to the individual nuclear cross sections were rigorously simplified for radiation propagating along special directions in the crystal [14] which are predictable from symmetry considerations. In these cases the incident γ-ray beam can be thought of as consisting of two orthogonal intensities each of which propagates through the crystal with its own index of refraction, and the analysis of experimental data to obtain internal magnetic fields and efg or msd parameters becomes more or less straightforward. It may be of interest to remark that even crystals belonging to cubic crystallographic space groups (such as yttrium iron garnet) may not be regarded as isotropic for Mössbauer experiments because of the possible Debye–Waller-factor anisotropies of the nuclear sites.

This paper has three main purposes: (1) To emphasize, by means of several examples, the importance of polarization effects in the analysis of data on single crystals or magnetized materials; (2) to further discuss symmetry restrictions on the index-of-refraction matrix of crystals and to outline an alternate derivation of the results of ref. [14] in this area; and (3) to point out that the simple equations referred to above are adequate for the analysis of data in many cases in which they are not rigorously correct.

The second general treatment, by Hannon and Trammell [15], of the intensity *versus* thickness problem mentioned above is very complete and elegant. They take into account boundary conditions, Bragg reflection, and Borrmann transmission in a treatment which is a generalization of the dynamical theory of X-ray diffraction. However, like Blume and Kistner, they have not been concerned with the inverse problem of deriving hyperfine or msd parameters from experimental data.

EXAMPLES ILLUSTRATING THE IMPORTANCE OF POLARIZATION

The most obvious effect of polarization is shown in a straightforward case by Fig. 1. Both spectra were taken with the same source and absorber and both have been corrected for the small ($\sim 10\%$) background. For the lower spectrum the enriched Fe absorber was demagnetized to as great an extent as possible prior to the measurement with a gradually decreasing 60-cycle field. The result was that for the most part there were several domains having different orientation within the thickness of the foil. For the upper spectrum a field of ~ 2 kOe was applied in the plane of the foil aligning most of the domains. It is easy to see that all the line depths and areas are decreased when the foil is magnetized. The explanation of this result is simple [16]. In the ideally unpolarized case (small randomly oriented domains) each line saturates with an effective cross section given by the product of a Clebsch–Gordan coefficient and an averaged angular intensity factor. For the magnetized absorber each individual absorption line is completely polarized. The unpolarized source radiation may be thought of as composed of two components of equal intensity, one having the polarization of the absorber line and one having an orthogonal polarization. The orthogonal component is transmitted without resonant absorption, while the parallel component is absorbed with a cross section determined from the normalization condition that for each polarization the sum of the cross sections for all the hyperfine components must be $nf'\sigma_0$, where n is the number of Fe^{57}

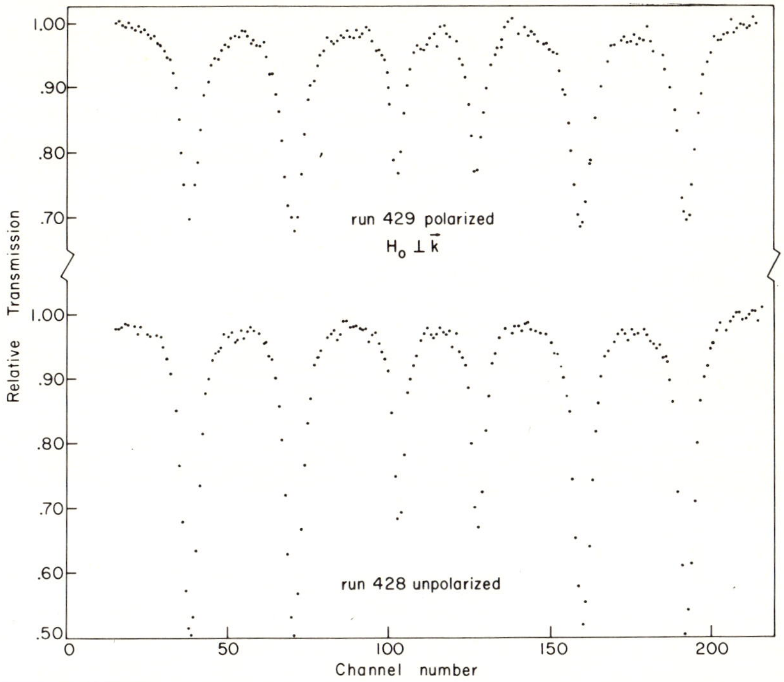

Fig. 1. Spectra taken with Co^{57} in Cu source and ~10 mg/cm² ~90% Fe^{57}-enriched Fe foil absorber at 23°C. (Unpublished data, R. M. Housley, 1964.)

atoms per unit area, f' is the Debye–Waller factor, and σ_0 is the fixed-nucleus cross section. With such a thick absorber the outer lines in the unpolarized case are absorbing considerably more than half of the recoil-free radiation coming from the source ($f = 0.71$ for Fe^{57} in Cu at 23°C [17]), and hence must decrease considerably in strength when the foil is magnetized.

Figures 2 and 3 illustrate another interesting effect of polarization in thick absorbers. Both spectra were taken with the same source and with absorbers from the same batch of $MgFe_2O_4$ enriched in Fe^{57}. In both cases the absorbers were in a longitudinal applied field of 55 kOe. In the spectrum of Fig. 2, which was taken with a relatively thin absorber, the outer pairs of lines corresponding to Fe in the octahedral [Fe] and tetrahedral (Fe) sites are clearly resolved [18]. The spectrum of Fig. 3, which was made with a much thicker sample of the same material, exhibits two unusual features: (1) There appear

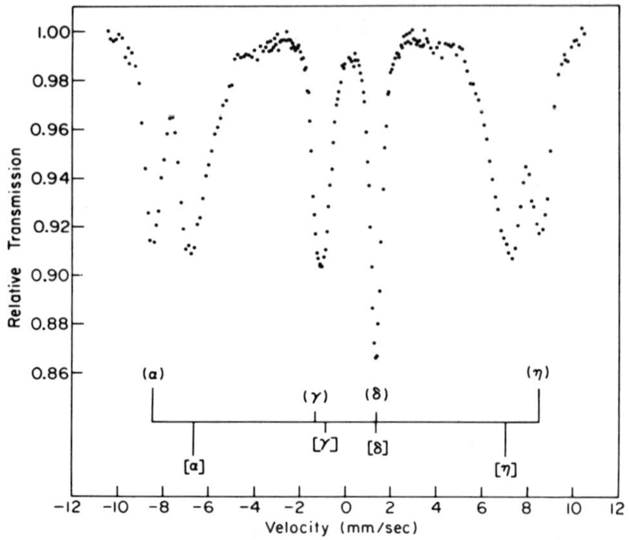

Fig. 2. Transmission spectra from a polycrystalline $MgFe_2O_4$ sample (2.8 mg/cm^2, 67.4% enriched in Fe^{57}) in a longitudinal magnetic field (H = 55 kOe) taken with a Co^{57}-Cu source. Source and absorber were at room temperature. The line positions of the tetrahedral (Fe) and octahedral [Fe] sublattices are indicated. (From ref. [18].)

to be extra strong lines between each pair of outer lines, and (2) the outer lines appear to be strongly thickness-broadened, but yet are not as deep as the narrower inner lines. This is especially clear for the left-hand octahedral line. Both of the peculiarities are due to polarization. The individual octahedral and tetrahedral lines are either completely right- or left-circularly polarized, and hence can absorb only half the recoil-free radiation from the source. The middle lines consist of overlapping octahedral and tetrahedral contributions, and hence can absorb all the incident recoil-free intensity. The same applies to the overlap region between the outer pairs of lines, so in a sufficiently thick absorber this region becomes more strongly absorbing than the positions of the line centers.

Figure 4 is a simple example of the effect of polarization on absorption intensities for a single crystal exhibiting a quadrupole doublet. The material $FeCO_3$ (siderite) is especially simple, having space group $R\bar{3}c$ with all Fe atoms located on axially symmetrical sites having their principal axes parallel or antiparallel. The curves of area ratio versus thickness were calculated assuming an unpolarized single line source [6]. The upper curve was calculated assuming the lines have, respectively, the maximum cross sections $3nf'\sigma_0/8$ and $5nf'\sigma_0/8$ and neglecting the effects of polarization. The lower curve was calculated taking into account that the weaker line only absorbs radiation polarized parallel to the z axis of the efg and that with a cross section $3nf'\sigma_0/4$, while the strong line absorbs radiation with that polarization with a cross section $nf'\sigma_0/4$ and radiation polarized perpendicular to z with a cross section $nf'\sigma_0$. It is interesting to note that because the $nf'\sigma_0/4$ contribution saturates most slowly, the thick-absorber area ratio actually falls below the value 3/5 characteristic of a thin absorber.

This siderite example was easy to analyze because the appropriate cross sections for two orthogonal component intensities were readily apparent. In low-symmetry crystals or crystals having resonant sites with efg tensors whose principal axes make oblique angles, or in

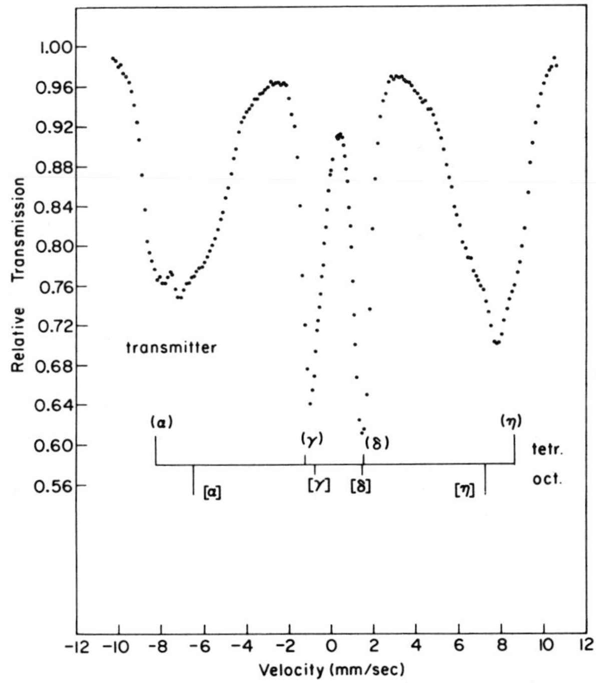

Fig. 3. As in Fig. 2, except for a thickness of 37.0 mg/cm², 67.4% enriched in Fe^{57}. (From ref. [18].)

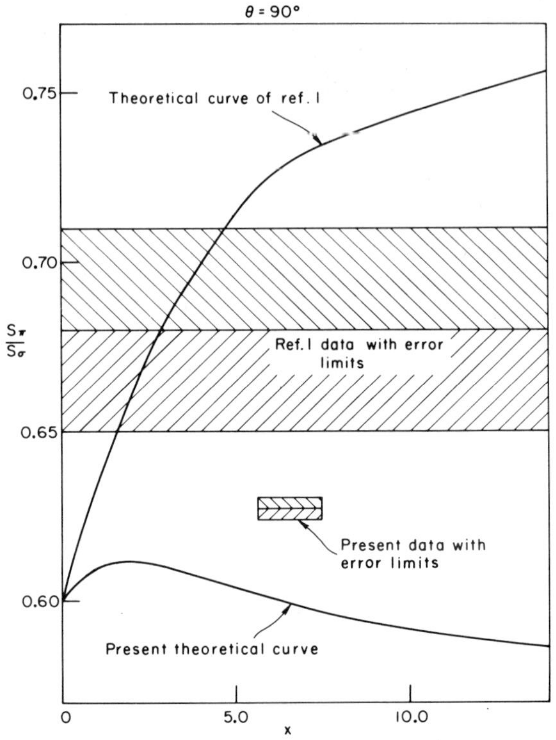

Fig. 4. Theoretical area ratios for γ-radiation propagating perpendicular to the c axis in a siderite crystal. The upper curve was obtained ignoring polarization effects, while the lower curve was obtained using the correct expression. (From ref. [6].)

crystals having internal magnetic fields which are noncollinear with the principal axes of the efg, it is not immediately apparent how this should be done. The discussion of symmetry restrictions in the next section shows how the orthogonal cross-section components may be found in terms of the unknown internal field parameters for observations along the appropriate symmetry directions.

DISCUSSION OF SYMMETRY RESTRICTIONS

Let us first consider the polarization of radiation emitted or absorbed with wave vector **k** due to a transition between a given pair of nuclear sublevels for a nucleus in a given type of site designated by *j*. For example, if we have axial symmetry at the nuclear sites and **k** is along the axis, then the radiation will have either right- or left-circular polarization. If **k** is perpendicular to the axis, then it will be linearly polarized along or perpendicular to the axis, assuming parity is conserved in electromagnetic transitions. For **k** at intermediate angles or for cases lacking axial symmetry the polarization will in general be elliptical. A number of examples have been presented graphically by Kistner [19]. Two possible ways of calculating these polarizations have been discussed in refs. [13] and [14].

The polarization of the cross section for the specified transition designated by i of the nucleus at site j may be described by a complex two-dimensional vector

$$\mathbf{A}_{ij} = a_1\hat{\mathbf{e}}_1 + a_2\hat{\mathbf{e}}_2 \tag{1}$$

where a_1 and a_2 are complex amplitudes and $\hat{\mathbf{e}}_1$ and $\hat{\mathbf{e}}_2$ are an arbitrary pair of orthogonal complex unit basis vectors. For example, $\hat{\mathbf{e}}_1$ could stand for right-circular polarization and $\hat{\mathbf{e}}_2$ for left-circular polarization. More familiarly, $\hat{\mathbf{e}}_1$ could be a unit vector along the x axis and $\hat{\mathbf{e}}_2$ a unit vector along the y axis.

Alternatively, the polarization of the cross section may be described in terms of a density matrix

$$\sigma^{ij} = \begin{pmatrix} \rho_{11}^{ij} & \rho_{12}^{ij} \\ \rho_{21}^{ij} & \rho_{22}^{ij} \end{pmatrix} \equiv \begin{pmatrix} a_1{}^*a_1 & a_1{}^*a_2 \\ a_2{}^*a_1 & a_2{}^*a_2 \end{pmatrix} \tag{2}$$

These two descriptions are equivalent, and one can easily also find the coefficients a_1 and a_2 if the density matrix is given. If, without loss of generality, we take a_1 real and positive, we have

$$a_1 = (\rho_{11}^{ij})^{\frac{1}{2}} \qquad \text{and} \qquad a_2 = \rho_{12}^{ij}/(\rho_{11}^{ij})^{\frac{1}{2}} \tag{3}$$

The coefficients a_1 and a_2 depend on the efg and internal magnetic field at the nuclear site, and through them so do the elements of the density matrix. We also see from the definition of σ^{ij} that

$$\rho_{11}^{ij}\rho_{22}^{ij} = \rho_{21}^{ij}\rho_{12}^{ij} \tag{4}$$

which is a relationship we will use later.

The intensity I and polarization of radiation incident on an absorber may also be described in terms of a 2×2 density matrix s. The reduction in intensity due to passage through a thin layer of nuclei of type designated by i and j will be proportional to the trace of the product $s \times \sigma^{ij}$. This fairly familiar result [20] is equivalent to saying that the reduction in intensity is proportional to the square of the absolute value of the projection of the vector representing the source polarization on the vector representing the absorption cross section and may be proved in that way, if the source radiation is completely polarized.

The electromagnetic interaction is sufficiently weak that an incident beam is attenuated by a negligible amount in passing through a layer of crystal of maximum unit-cell dimensions. Therefore, the reduction in intensity for each pair of i and j is given by a similar expression. Since the distributive property holds for matrix multiplication and addition, it follows that absorption by the whole crystal can be described by a density matrix which is obtained by summing the individual cross-section matrices over i and j. The density matrix for a fine, randomly oriented powder in a magnetic field may be obtained by making a properly weighted average of the density matrix over all orientations of the grains. From what has just been said we see that

$$\sigma = \begin{pmatrix} \sigma_{11} & \sigma_{12} \\ \sigma_{21} & \sigma_{22} \end{pmatrix} \equiv \sigma_0 \sum_{ij} \begin{pmatrix} \rho_{11}^{ij} & \rho_{12}^{ij} \\ \rho_{21}^{ij} & \rho_{22}^{ij} \end{pmatrix} \frac{n_j f_j'}{(x-x_{ij})^2+1} \tag{5}$$

where σ_0 is the fixed-nucleus cross section, n_j the number of sites of type j per unit area, f_j' the Debye–Waller factor for site j, and $x = 2E/\Gamma$ is two times the energy divided by the natural linewidth. In general, if lines are broadened, their polarizations will not be accurately known; hence, for purposes of this analysis natural linewidth is assumed.

Now, for **k** along special directions in a crystal there exist groups of symmetry operations—reflections, rotations, inversion, or combinations of these with time reversal—which leave both **k** and the crystal invariant. If a given symmetry operation has transformed site j into site j', then the polarization of the cross section due to the nucleus at j' will be given by

$$\mathbf{A}_{ij'} = a_1\hat{\mathbf{e}}_1' + a_2\hat{\mathbf{e}}_2' \tag{6}$$

where $\hat{\mathbf{e}}_1$ and $\hat{\mathbf{e}}_2$ have been transformed to $\hat{\mathbf{e}}_1'$ and $\hat{\mathbf{e}}_2'$, respectively. The new basis vectors will be simply related to the old, so it is easy to express the polarization of the cross section of site j' in the same basis system as that of site j.

As an example, assume **k** lies in a mirror plane, and choose $\hat{\mathbf{e}}_1$ normal and $\hat{\mathbf{e}}_2$ parallel to the plane. Reflection in the plane leaves $\hat{\mathbf{e}}_2$

invariant and changes $\hat{\mathbf{e}}_1$ to $-\hat{\mathbf{e}}_1$. Therefore, the polarization at site j' may be written as

$$\mathbf{A}_{ij'} = -a_1\hat{\mathbf{e}}_1 + a_2\hat{\mathbf{e}}_2 \tag{7}$$

Forming the density matrices as in Eq. (2), we see

$$\sigma^{ij} = \begin{pmatrix} a_1{}^*a_1 & a_1{}^*a_2 \\ a_2{}^*a_1 & a_2{}^*a_2 \end{pmatrix} \quad \text{and} \quad \sigma^{ij'} = \begin{pmatrix} a_1{}^*a_1 & -a_1{}^*a_2 \\ -a_2{}^*a_1 & a_2{}^*a_2 \end{pmatrix} \tag{8}$$

hence the sum of these contributions to σ is diagonal, and the same applies to any other pair of sites related by reflection in this plane. Therefore, σ is diagonal in this basis system. Since reflection and inversion change right-handed to left-handed coordinate systems, they always change the sense of rotation for elliptically polarized radiation.

If we further restrict $\mathbf{k}$ to lie along a fourfold rotation axis in a mirror plane, we find that σ must be a multiple of the unit matrix.

For any real crystal it is a simple matter to find all the symmetry directions in which σ may be written in a diagonal form without first knowing the internal field, efg and msd parameters. These are also just the directions in which the dispersive part of the index of refraction must commute with the absorptive part (being diagonal in the same basis system) and hence does not influence the absorption intensities [13, 14, 18].

The polarization and intensity I of the source may be described by a density matrix

$$s = \frac{I}{\pi}\begin{pmatrix} s_{11} & s_{12} \\ s_{21} & s_{22} \end{pmatrix} \equiv \frac{I}{2\pi}\sum_{k,l}\begin{pmatrix} \rho_{11}^{kl} & \rho_{12}^{kl} \\ \rho_{21}^{kl} & \rho_{22}^{kl} \end{pmatrix} \frac{r_l f_l}{(x - x_{kl} - v)^2 + 1} \tag{9}$$

where r_l is the fraction of source atoms in site l, f_l is the Debye–Waller factor for the site, and v is the normalized Doppler shift. We have without loss of generality assumed that the source moves and the absorber is held fixed.

Assuming we choose our basis system so that σ is diagonal, the intensity transmitted through the crystal at any energy x and Doppler shift v is

$$I(x,v) = (I/\pi)[s_{11}e^{-\sigma_{11}} + s_{22}e^{-\sigma_{22}}] \tag{10}$$

This only differs from the expression for the unpolarized case in having two terms on the right instead of one. Therefore, formulas for depths [10] and areas [11] of lines will differ from the corresponding unpolarized formulas only by consisting of two separate contributions, one for each orthogonal polarization. Some useful approximate formulas and series expansions have been given in ref. [14].

APPROXIMATE TREATMENT FOR LOWER SYMMETRY DIRECTIONS

In many cases where the dispersive part of the index of refraction cannot rigorously be neglected it plays such a minor role that it may be ignored to a good approximation. Roughly, these cases involve only fairly well-resolved lines, and absorbers which are not extremely thick (e.g., are not made with separated isotopes).

In these cases, although the basis in which the cross-section matrix will be diagonal cannot be chosen in advance, it may be found either by using a polarized source or by using an unpolarized source and analyzing the polarization of the transmitted radiation (suggested by J. P. Hannon).

With an unpolarized source, even without analysis of the transmitted beam, the intensities may be related to the internal field and msd parameters if measurements are made on two samples of different thickness. For an unpolarized source Eq. (10) becomes

$$I(x, v) = (I/2\pi)[e^{-\sigma_{11}} + e^{-\sigma_{22}}] \qquad (11)$$

and it is apparent that the transmitted intensity is determined by the average cross section $p = (\sigma_{11} + \sigma_{22})/2$ and the fractional polarization $a = (\sigma_{11} - \sigma_{22})/(\sigma_{11} + \sigma_{22})$. From measurements on samples of two different thicknesses, both these quantities may be determined. The quantities p and a are independent of the basis system in which σ is described and may be related to the internal field and msd parameters. In an arbitrary basis [13]

$$p = \frac{\sigma_{11} + \sigma_{22}}{2} \quad \text{and} \quad a = \frac{[(\sigma_{11} - \sigma_{22})^2 + 4\sigma_{12}\sigma_{21}]^{\frac{1}{2}}}{\sigma_{11} + \sigma_{22}} \qquad (12)$$

Simple formulas for extracting p and a from data on relatively thin absorbers are given in ref. [14]. For a sufficiently thin absorber (frequently not practical to prepare) the absorption is independent of a.

Since no exact criterion for neglecting the dispersive part of the index of refraction has been given, I suggest inserting the index of refraction [13, 14, 18] obtained with parameters determined from the analysis outlined above into Eq. (5″) of ref. [13] and calculating theoretical absorption intensities. If they agree within statistics to the measured intensities, then neglect of the dispersive part was justified.

ACKNOWLEDGMENTS

I have profited from many discussions of this work with U. Gonser and R. W. Grant. I would like to thank M. Blume and J. P. Hannon for sending me preprints of refs. [13] and [15], and also acknowledge helpful discussions with them.

REFERENCES

1. P. Zory, *Phys. Rev.* **140**:A1401 (1965).
2. J. Danon and L. Iannarella, *J. Chem. Phys.* **47**:382 (1967).
3. R. Ingalls, K. Ono, and L. Chandler, *Phys. Rev.* **172**:295 (1968).
4. K. Chandra and S. P. Puri, *Phys. Rev.* **169**:272 (1968).
5. R. W. Grant, R. M. Housley, and U. Gonser, *Phys. Rev.* **178**:523 (1969).
6. R. M. Housley, U. Gonser, and R. W. Grant, *Phys. Rev. Letters* **20**:1279. (1968).
7. U. Gonser, R. W. Grant, H. Wiedersich, and S. Geller, *Appl. Phys. Letters* **9**:18 (1966).
8. U. Gonser and R. W. Grant, *Phys. Stat. Sol.* **21**:331 (1967).
9. R. W. Grant, S. Geller, H. Wiedersich, U. Gonser, and L. D. Fullmer, *J. Appl. Phys.* **39**:1122 (1968).
10. J. G. Dash, R. D. Taylor, D. E. Nagle, P. P. Craig, and W. M. Visscher, *Phys. Rev.* **122**:1116 (1961).
11. G. A. Bykov and Pham Zuy Hien, *Zh. Eksperim. i Teor. Fiz.* **43**:909 (1962) [*Soviet Phys.—JETP* **16**:646 (1963)].
12. S. L. Ruby and P. A. Flinn, *Rev. Mod. Phys.* **36**:351 (1964).
13. M. Blume and O. C. Kistner, *Phys. Rev.* **171**:417 (1968).
14. R. M. Housley, R. W. Grant, and U. Gonser, *Phys. Rev.* **178**:514 (1969).
15. J. P. Hannon and G. T. Trammell, submitted to *Phys. Rev.*
16. R. H. Nussbaum and R. M. Housley, *Nucl. Phys.* **68**:145 (1965).
17. R. M. Housley, J. G. Dash, and R. H. Nussbaum, *Phys. Rev.* **136**:464 (1964).
18. R. M. Housley and U. Gonser, *Phys. Rev.* **171**:480 (1968).
19. O. C. Kistner, in: *Hyperfine Structure and Nuclear Radiations*, E. Matthias and D. A. Shirley, eds. (North-Holland Publishing Co., Amsterdam, 1968), p. 295.
20. U. Fano, *Rev. Mod. Phys.* **29**:74 (1957).

Determination of Zero-Point Phonon Parameters: Calibration of the Second-Order Doppler Shift*

T. A. Kitchens and P. P. Craig

Brookhaven National Laboratory
Upton, New York

and

R. D. Taylor

Los Alamos Scientific Laboratory
Los Alamos, New Mexico

A considerable body of research has been devoted to the determination of the Mössbauer recoil-free fraction, or f*-factor, from which the mean square atomic displacement of resonant nuclei can be deduced. A quantity intimately related to the mean square displacement, but considerably less accessible to experiment, is the mean square velocity. Normally this quantity is hidden amid a wealth of complexity relating to the chemical shifts of source and absorber. We have developed a technique of intercomparing two quantities, both directly determined from the Mössbauer experiments—the recoil-free fraction and the second order Doppler shift—so as to permit deduction of the absolute zero-point mean-square velocity. This quantity has not been hitherto determined in any experiments of which we are aware, and we believe it will prove valuable in permitting tests of theoretical models of impurities in crystal lattices. Results are presented for a number of systems involving both the* Fe^{57} *and the* Sn^{119} *resonances.*

INTRODUCTION

The determination of the chemical properties of resonant nuclei has been among the most significant uses of the Mössbauer effect. One

*Work performed under the auspices of the US Atomic Energy Commission.

of the important parameters emerging from Mössbauer experiments in chemical studies is the isomeric shift, a shift which depends on the local environs of the resonant nuclei. This isomeric shift is not the only source of shifts in the mean velocity of the Mössbauer spectrum; the second-order Doppler mechanism also produces a major contribution. In this paper a technique is described by which the absolute root-mean-square zero-point motion (ZP velocity) can be determined. This technique also provides a method of eliminating the effects of the second-order Doppler mechanism more rigorously than heretofore possible, and thereby removes a source of uncertainty in the isomeric shift. The technique also allows the observed ZP velocity to be related to models in much the same way as the recoil-free fractions have been.

Recent work on the mechanisms controlling the critical temperature in superconductivity shows that a parameter likely to be of great importance is related to the ratio of the expectation of the mean-square velocity to the mean-square displacement within the lattice near zero temperature [1]. The method introduced in this paper can be used to measure this ratio. It may therefore be hoped that application of the techniques presented here will eventually prove useful in understanding some aspects of superconductivity.

Many of the experimental results and portions of the analysis have been reported previously [2] but the discussion of the effect of mass impurities, the connection with superconductivity, and the data for dilute Fe in Mo are new.

METHOD

The mean energy of the radiation emitted from a Mössbauer source may be written

$$E = E_0[1 + (v/c)] - (\Delta E)_{\text{iso}} - (\Delta E)_{2D} \tag{1}$$

E being the emitted photon energy and v the Doppler velocity. The isomeric shift, the shift in the energy due to the s-electron density at the nucleus, is denoted by $(\Delta E)_{\text{iso}}$, and $(\Delta E)_{2D}$ is the shift arising from the second-order Doppler shift. Details and the systematics of the isomeric shift, sometimes called the chemical shift, have been discussed elsewhere [3]. The second-order Doppler shift can equivalently be considered as arising from the change in the mass of the emitting lattice due to the loss of a photon having a relativistic mass equivalent to E/c^2. This argument was first presented by Josephson [4], and the shift is often labeled the Josephson shift.

In addition to the factors discussed above, there are various sources of broadening and splitting the gamma-ray energies, such as an electric field gradient or a magnetic field, but these effects do not change the mean gamma-ray energy [3]. An exception to this statement occurs when the splitting is large enough (of the order kT) to affect the nuclear energy-level populations. At temperatures below 0.4°K such effects have been observed.

Of the terms in Eq. (1) only the last is normally considered to be temperature dependent. (Temperature dependence of the isomeric shift can be produced by changes in phase, lattice expansion, and other mechanisms which change the s-electron density at the nucleus. These small effects will be ignored here.) We will describe a technique to determine the value of $(\Delta E)_{2D}$ at absolute zero. Then an accurate value of $(\Delta E)_{\text{iso}}$ can be determined using Eq. (1).

Restricting attention to the isotropic systems and for the moment neglecting localized modes, $f(T)$, the recoil-free fraction at the temperature T, and $(\Delta E)_{2D}$ are given by

$$f(T) = \exp(-\langle x^2\rangle_{\text{av}}/\lambda\!\!\!^{-2}) \tag{2}$$

and

$$(\Delta E)_{2D} = -E\langle v^2\rangle_{\text{av}}/2c^2 \tag{3}$$

Here $\langle x^2\rangle_{\text{av}}$ is the mean-square displacement of the emitting nucleus along the direction of the gamma-ray emission; $\langle v^2\rangle_{\text{av}}$ is the mean-square velocity of the nucleus. The velocity of light is denoted by c, and $\lambda\!\!\!^{-} = \hbar c/E$ is the reduced wavelength of the gamma radiation having an energy E.

Rewriting Eq. (1) in terms which have more meaning experimentally and separating the ZP velocity and the temperature-dependent contribution to the second-order Doppler term,

$$\frac{\Delta E}{E} = \left(\frac{\delta v}{c}\right)_T + \left(\frac{\delta v}{c}\right)_{\text{ZP}} + \left[\left(\frac{\delta v}{c}\right)_{\text{iso}} + \left(\frac{\delta v}{c}\right)_{\text{abs}}\right] \tag{4}$$

where the δv is the first-order Doppler shift necessary to compensate for the energy shift due to the mechanism indicated by the subscript. The subscript T indicates the temperature-dependent second-order Doppler contribution. The last term is from similar energy shifts in the absorber and will be independent of the temperature of the source. Using Eqs. (2) and (3), we can rewrite Eq. (4) as

$$\ln f(T) = \frac{2c^2\langle x^2\rangle_{\text{av}}}{\lambda\!\!\!^{-2}\langle v^2\rangle_{\text{av}}}\left\{\frac{\Delta E}{E} - \left[\left(\frac{\delta v}{c}\right)_{\text{iso}} + \left(\frac{\delta v}{c}\right)_{\text{abs}}\right]\right\} \tag{5}$$

The terms in square brackets are independent of the source temperature. The recoil-free fraction $f(T)$ and the shift $\Delta E/E$ are directly measurable. At high temperatures it turns out that $[\langle x^2\rangle_{av}/\langle v^2\rangle_{av}]$ is nearly independent of the temperature. Hence the slope of $\ln f(T)$ *versus* $\Delta E/E$ determines $\langle x^2\rangle_{av}/\langle v^2\rangle_{av}$ at high temperatures. Furthermore, $[\langle x^2\rangle_{av}/\langle v^2\rangle_{av}]_{ZP}$ can be determined. This quantity and $f(0)$, the recoil-free fraction at absolute zero, are then used to calculate $(\delta v/c)_{ZP}$.

The procedure just outlined would have little utility unless the determination of $[\langle x^2\rangle_{av}/\langle v^2\rangle_{av}]$ is feasible. The feasibility of this determination will be demonstrated in the following section for some experimental data. Some theoretical justification for this result and the detailed procedures for determining $[\langle x^2\rangle_{av}/\langle v^2\rangle_{av}]_{ZP}$ and $(\delta v/c)_{ZP}$ will be presented below.

EXPERIMENTAL RESULTS

In order to characterize the results of experiment in as concise a fashion as possible, we define the coefficient $S(T)$. That is, we write

$$\ln f(T) \equiv S(T)\left\{\frac{\Delta E}{E} - \left[\left(\frac{\delta v}{c}\right)_{\text{iso}} + \left(\frac{\delta v}{c}\right)_{\text{abs}}\right]\right\} \tag{6}$$

The power of our method is that $S(T)$ turns out to be essentially temperature-independent, so that if $f(0)$ is known (and techniques for finding this quantity are highly developed) then $S(0)$ and $(\delta v/c)_{ZP}$ can be obtained and related to various lattice models.

Our procedure is to plot experimentally determined values of $\ln f(T)$ against corresponding values of the shift $\Delta E/E$. The f values used are the *absolute* f values, and the observed Doppler shifts have been converted to relative γ-ray energy shifts to avoid the confusion of the sign of the Doppler velocity necessary to restore resonance when comparing both absorber and source experiments. In this type of presentation, temperature is a parameter, and only quantities directly determined from Mössbauer effect experiments need be considered.

In Fig. 1 we present such a plot for the 23.9-keV M.E. resonance in a SnTe absorber. The source was Sn^{119m} in SnTe held at 19.4°K; since neither f nor $\Delta E/E$ varies much with temperature below about 30°K, errors due to temperature drift in the source were minimized. A rough temperature scale for the data is given. The linearity of the plot is apparent, except for a small deviation at the lowest temperatures.

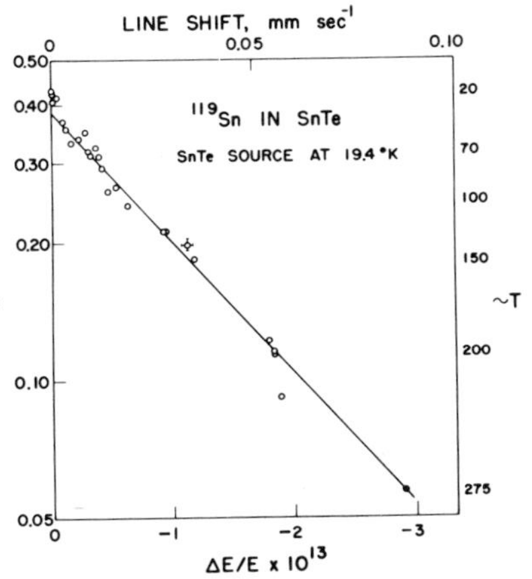

Fig. 1. Recoil-free fraction *f versus* the energy shift of the 23.9-keV γ-ray of Sn^{119} in a SnTe absorber. The Doppler shift to restore resonance is given at the top, and an approximate temperature scale is given on the right side. These data, taken from refs. [2, 15], have typical random errors of the magnitude shown near the center of the figure.

From $S(T)$ and $f(0)$ we show below that we can determine the absolute value of the mean-square zero-point velocity.

Figure 2 includes some results obtained by Steyert and Taylor [5] for dilute solutions of Fe^{57} in several cubic metallic hosts. The $f(T)$ for these sources were obtained using an enriched FeTi broad absorber at room temperature, and the shifts were obtained with a metallic Fe absorber. Temperatures ranged from 4 to 1020°K. On each graph the point at room temperature is indicated by a solid circle.

The data presented here are representative. A number of similar plots may be found in ref. [2]. All systems for which precise experimental data are available show similar linearity. Unfortunately, the

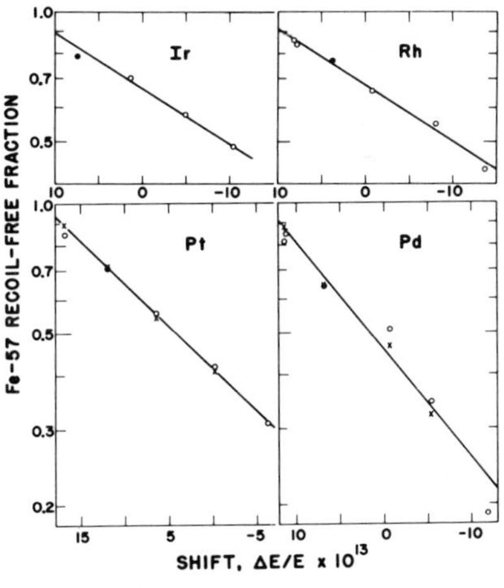

Fig. 2. Plots of $\ln f$ *versus* the energy shift of the 14.4-keV γ-ray for dilute Fe^{57} in various hosts for temperatures between 4 and 1020°K. Room-temperature data are denoted by a solid circle. The shift is measured relative to a Fe^{57} absorber at 298°K. Additional information is given in ref. [2]. Data designated with crosses are from $f(T)$ measurements from ref. [14].

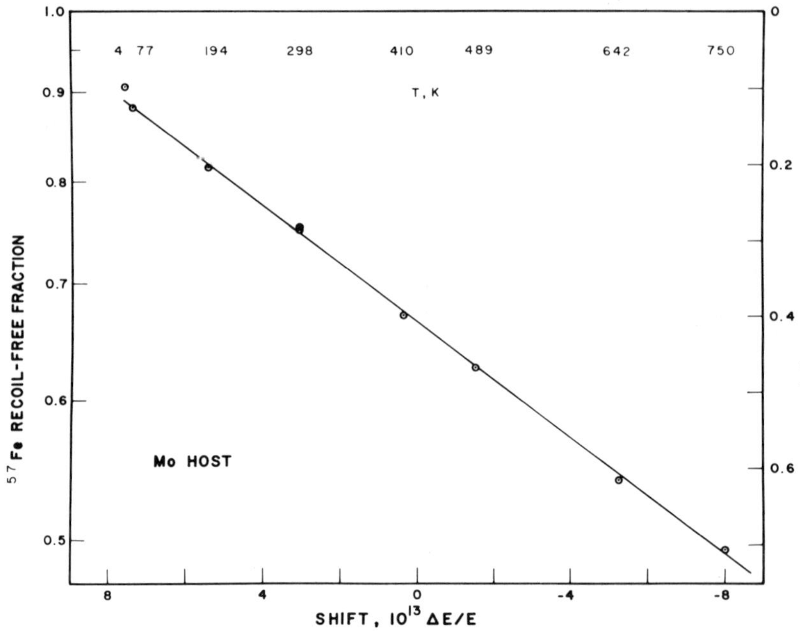

Fig. 3. Recoil-free fraction f *versus* the γ-ray energy shift of the 14.4-keV γ-ray of dilute Fe^{57} in Mo. An approximate temperature scale is given at the top. The shift $(\Delta E/E)$ is relative to a 298°K potassium ferrocyanide absorber and has a precision of $\pm 0.05 \times 10^{-13}$. The precision of $\ln f$ is ± 0.005.

number of such carefully studied systems is thus far rather restricted. Data for dilute Fe in Mo are given in Fig. 3 and will be discussed below.

We now turn to the significance of the linearity of plots of $\ln f$ *vs.* shift, and to the problem of extracting solid-state information from the data.

MODELS FOR $S(T)$

Homogeneous Lattice

In the harmonic approximation the mean-square displacement of an atom of mass m in an isotropic lattice is

$$\langle x^2 \rangle_{av} = (\hbar/m)\langle \omega^{-1} \rangle \tag{7}$$

The mean-square velocity of the atom is

$$\langle v^2 \rangle_{av} = 3\langle v_x^2 \rangle_{av} = 3(\hbar/m)\langle \omega \rangle \tag{8}$$

where we have denoted the thermal average of a power of ω as

$$\langle \omega^l \rangle \equiv \int_0^\infty g(\omega)(\bar{n} + \tfrac{1}{2})\omega^l \, d\omega \tag{9}$$

and the occupation number is $\bar{n} = 1/[\exp(\hbar\omega/kT) - 1]$. The density of states $g(\omega)$ is taken normalized to the N degrees of freedom.

Thus, from Eqs. (5)–(8) one may write

$$S(T) = \frac{2c^2}{\lambda^2} \frac{\langle x^2 \rangle_{\text{av}}}{\langle v^2 \rangle_{\text{av}}} = \frac{2c^2}{3\lambda^2} \frac{\langle \omega^{-1} \rangle}{\langle \omega \rangle} \tag{10}$$

Details of the evaluation of $S(T)$ for the Debye model and the Einstein model, and for $g(\omega)$ determined for metallic Fe by neutrons, are given in ref. [2]. We will only summarize the results here. On the Debye model $S(T)$ is approximately constant for temperatures $T \gtrsim \frac{1}{2}\,\theta_D$, while at lower temperatures $S(T)$ decreases by a factor of 2/3, i.e., $S(0) = \frac{2}{3}\,S(\infty)$. The limit of the $S(T)$ at low temperature for the $g(\omega)$ for metallic Fe is 0.699 of the high temperature $S(\infty)$ rather than 0.667 given by the Debye model, as shown in Fig. 4. The Einstein model predicts the same temperature dependence for $\langle v^2 \rangle_{\text{av}}$ and $\langle x^2 \rangle_{\text{av}}$, so $S(T)$ is independent of temperature.

A number of parameters derivable from the observed values of $S(T)$ are presented in Table I. A more complete set is found in Table I of ref. [2].

In particular, a characteristic temperature ψ_0 defined by

$$(k\psi_0)^2 = \hbar^2[\langle \omega \rangle / \langle \omega^{-1} \rangle]_{\text{ZP}}$$

provides a measure of the ratio $[\langle \omega \rangle / \langle \omega^{-1} \rangle]_{\text{ZP}}$ which arises in McMillan's treatment of the critical temperature of strongly coupled superconductors [1]. In fact, McMillan has shown that it is this phonon parameter which governs the critical temperature for the $5d$ transition-metal series superconductors rather than the electronic properties. Here we see that this quantity is obtained rather directly from Mössbauer measurements. We will further consider this correlation between the Mössbauer effect and superconductivity later.

Lattices with Impurities

Almost invariably, a systematic Mössbauer investigation involves attempting to learn something about a lattice by observing the properties of an impurity, the Mössbauer atom.

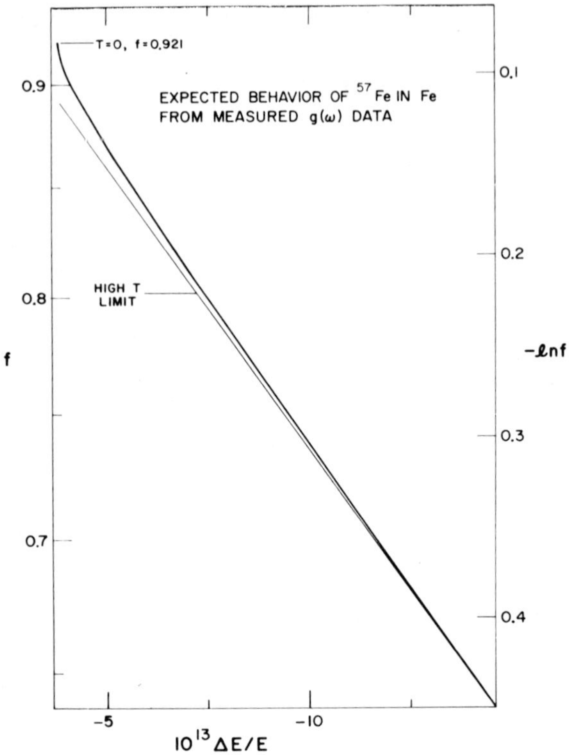

Fig. 4. The theoretical $\ln f$ *versus* γ-ray energy shift calculated from the $g(\omega)$ determined by the 300°K inelastic neutron-scattering data of ref. [17]. The range shown corresponds to 0–590°K.

The effects on the lattice dynamics of a solid due to the presence of an impurity involve at least (1) the change in mass, and (2) the difference in the potential at the impurity site from that at the sites of the host atoms [6–8]. It is more difficult to consider the second factor, even neglecting the complications of anharmonicity introduced by the impurity [9]. Due to a lack of informative experimental data on this subject, the validity of the existing theories has not been checked. For this reason we will restrict our attention to the soluble situation, where the only effect of the impurity is the change in mass.

In case the Mössbauer atom has a mass significantly different from the host atom, localized modes do become important, and the formalism of the type developed by Visscher [6] or by Elliott [8] may be expected to modify the simple picture of the homogeneous lattice.

Table I

Parameter*	Sn^{119} SnTe	Sn^{119} Nb_3Sn	Fe^{57} Ir	Fe^{57} Rh	Fe^{57} Pt	Fe^{57} Pd	Fe^{57} Au	Fe^{57} Cu	Fe^{57} Ti	Fe^{57} Fe†	Fe^{57} Mo
θ_D, host,‡ °K	—	228	420	478	233	275	165	335	430	(413)	460
$\theta_D(M/M')^{\frac{1}{2}}$, °K	—	—	771	642	431	376	307	354	394	413	496
$f(0)$	0.43	0.44	(0.92)	0.87	0.91	0.90	0.87	0.91	0.82	0.921	0.906
$-(\delta v/c)_0 \times 10^{13}$	2.0	5.1	4.3	6.9	3.5	3.3	2.8	2.6	3.9	3.76	3.85
ψ_0, °K	108	178	311	305	249	221	193	215	191	292	271
$f(0)$ [from Eq. (13)]	0.75	0.84	0.93	0.92	0.91	0.90	0.88	0.89	0.88	0.921	0.915
$x_{ZP} \times 10^8$, cm	0.076	0.075	0.040	0.051	0.042	0.044	0.051	0.042	0.061	0.0393	0.043

*Definition equations are given in the text, with the exception of $x_{ZP} = \langle x_0^2 \rangle^{\frac{1}{2}} = \lambda\!\!\!^{-}[-\ln f(0)]^{\frac{1}{2}}$. All quantities except θ_D are deduced from Mössbauer effect measurements.

†Fe^{57} in Debye solid with $m_{host} = 57$, $\theta_D = 413$°K.

‡Values from Bijl [18], except that for Nb_3Sn, which is from Vieland and Wicklund [19], and that for Mo, which is from McMillan [1].

Visscher has shown that in many circumstances a good first approximation is to use the Debye model, for which $g(\omega)$ is proportional to ω^2, and use an effective Debye temperature given by $\theta_D(M/M')^{\frac{1}{2}}$, where M is the host atomic mass and M' is the Mössbauer atom's mass.

Elliott has developed a formalism exact for an arbitrary mass impurity in anharmonic lattice. He gives an integral equation for $\langle x^2\rangle_{\text{av}}$ and $\langle v^2\rangle_{\text{av}}$ for the cubic lattice with arbitrary $g(\omega)$. The impurity (mass M') was considered as a mass defect $\epsilon = (M - M')/M$. Within the Debye model one always finds a localized mode above the Debye cutoff frequency ω_D for a light mass impurity, $\epsilon > 0$. For heavy mass impurities, $\epsilon < 0$, there are frequencies within the continuum which have a local maximum in the effective density of states.

The temperature dependences predicted from the Elliott model for $\langle x^2\rangle$ and for $\langle v^2\rangle$ are qualitatively similar to those found from the simple Debye model. We have examined the ratio $[\langle v^2\rangle/\langle x^2\rangle] = 3[\langle\omega\rangle/\langle\omega^{-1}\rangle]$ for various mass ratios R. The results are plotted *versus* reduced temperature in Fig. 5. The temperature scale is reduced according to the ideas of Visscher [6], and the success of this procedure is seen in the fact that the inflection points for all curves occur near the same value of the reduced temperature.

Figure 6 is a further summary of calculations made on the mass-impurity Debye model. Here a quantity related to the $S(T)$ by Eq. (10), $2R[\langle\omega\rangle/\langle\omega^{-1}\rangle]/\omega_D{}^2$, is plotted *versus* $R = M'/M$ in the limits of high and low temperatures. It is quite clear from Fig. 6 that $S(T)$ is sensitive

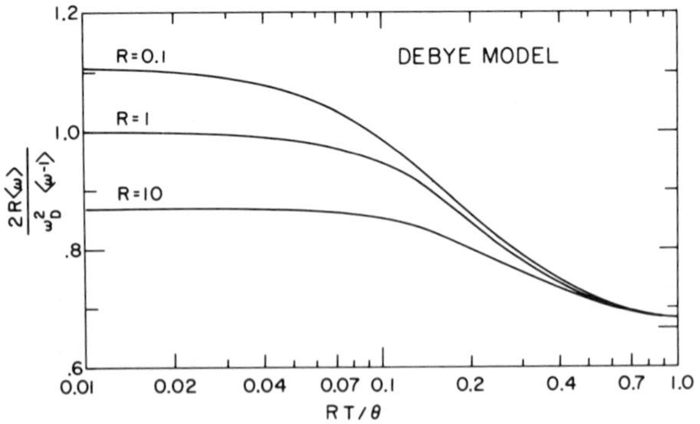

Fig. 5. The temperature dependence of $[\langle\omega\rangle/\langle\omega^{-1}\rangle] = \frac{2}{3}(c/\lambda\!\!\!^-)^2/S(T)$ as a function of the reduced temperature $(M'/M)^{\frac{1}{2}}T/\theta$ for various mass ratios $R = M'/M$ for a mass-defect isotropic Debye model. The high-temperature value of $S(T) = S(\infty)$ is given by the line on the right immediately below the $S(T)$ curves.

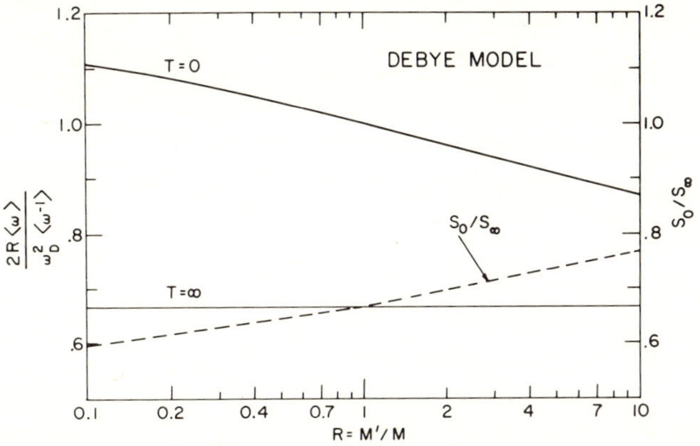

Fig. 6. The high- and low-temperature asymptotic values of $[\langle\omega\rangle/\langle\omega^{-1}\rangle]$ and the ratio of these values (dotted line) *versus* the ratio R of the impurity mass to the host mass.

to R, but that the scaling of the Debye θ by $1/\sqrt{R}$ suggested by Visscher [6] removes most of the R dependence. The ratio $S(\text{high } T)/S(0)$, equal to 1.5 for the homogeneous Debye model, is also sensitive to R, but is between 1.6 and 1.3 for $0.25 < R < 8$.

The Zero-Point Second-Order Doppler Shift

As mentioned above, the zero-point motion produces a second-order Doppler shift in the γ-ray energy, $(\delta v/c)_{ZP}$. For molecular solids it has been pointed out by Hazony [10] that this temperature-independent term is comparable to the isomer shifts. Hazony [10] and also Clark *et al.* [11] have used superimposed Einstein models to estimate $(\delta v/c)_{ZP}$ for various iron compounds from high-temperature measurements of $(\delta v/c)_T$. Here we suggest a method appropriate for the isotropic mass-defect Debye model based on the calculations of the previous subsection. The method is analogous to that developed for the homogeneous Debye and Einstein models discussed in ref. [2].

Because the high-temperature $S(T) = S(\infty)$ is constant, it is possible to determine it from the slope of the $\ln f(T)$ *versus* $\Delta E/E$ plot. In fact, as is pointed out in Appendix A, it is not even necessary to make absolute measurements of $f(T)$ to determine $S(\infty)$. Within the mass-defect Debye model $S(0)$ can be determined from the measured $S(\infty)$, R, and Fig. 6. Then $(\delta v/c)_{ZP}$ can be calculated from the absolute recoil-free fraction $f(0)$ by

$$(\delta v/c)_{ZP} = [\ln f(0)]/S(0) \tag{11}$$

As pointed out before, techniques for determining $f(0)$ are highly developed, and the mass-defect calculations suggest that, within a few per cent, $S(0) = S(\infty)/1.4$ for most experimental conditions (see Fig. 6).

In Table I the $(\delta v/c)_{ZP}$ values were calculated using $S(0) = S(\infty)/1.5$, the homogeneous Debye model result. The above method would reduce these values, but not by more than 15%.

Application to Superconductivity

From the results of McMillan's work it appears that a decrease in $[\langle\omega\rangle/\langle\omega^{-1}\rangle]_{ZP}$ usually corresponds to an increase in the critical temperature of the superconductor. From Fig. 6 it is apparent that, at least locally, $[\langle\omega\rangle/\langle\omega^{-1}\rangle]_{ZP}$ is decreased by the addition of a heavy impurity to a light host lattice. A detailed understanding of the superconductivity would depend upon the superconducting coherence length ξ and upon the physical extent of the parameter $[\langle\omega\rangle/\langle\omega^{-1}\rangle]$ associated with the heavy impurity. Although no quantitative estimate of the extent has been made, Elliott has suggested that it increases as ϵ (*mass defect*) decreases in magnitude. Certainly, the theory as mentioned above is inadequate for most real lattices, but we believe that the qualitative result is correct. Curves similar to those in Fig. 6 for more-realistic models could be used to obtain $[\langle\omega\rangle/\langle\omega^{-1}\rangle]_{ZP}$ for the host lattice from $[\langle\omega\rangle/\langle\omega^{-1}\rangle]_{ZP}$ measured by the Mössbauer effect for an impurity. Such calculations are in progress.

Nonetheless, the Mössbauer technique should prove capable of obtaining unique information on the behavior of $[\langle\omega\rangle/\langle\omega^{-1}\rangle]_{ZP}$ in real systems. This is, perhaps, the first significant correlation of the phenomenon of superconductivity and the Mössbauer technique [12].

KAGAN'S INTEGRAL RELATION

A method of analysis supplemental to that given above has been proposed by Kagan [13].* He finds that for a harmonic system the temperature-dependent part of the second-order Doppler shift is related to the zero-temperature recoil-free fraction,

$$\ln f(0) = \frac{2}{\pi^2}\frac{E^2}{k^2}\int_0^\infty \left(\frac{\partial v}{c}\right)_T \frac{dT}{T^3} \tag{12}$$

*The $|\ln f|$ value predicted by his Eq. (3.8) is a factor of two too high.

The term $(\delta v/c)_T$ is the temperature-dependent part of the second-order Doppler-shift contribution to the observed shift.

We have evaluated this integral for several systems assuming any other temperature-dependent shifts to be small or self-compensating. The results are summarized in Table II. The low-temperature contribution to the integral in Eq. (12) is critically dependent on the choice

Table II. Comparison of $f(0)$ from Different Techniques

System	$f(0)$, (Kagan integral)*	$f(0)$, [Eq. (13)]†	f, measured‡	T, °K
Fe^{57} in Cu	0.90_5	0.89	0.910 ± 0.002§	4
			(0.913)	0
Fe^{57} in Pd	0.91_3	0.90	0.891 ± 0.002	19
			0.875 ± 0.002	13
			(0.896)	0
Fe^{57} in Pt	0.89_9	0.91	0.905 ± 0.003	20
			0.897 ± 0.003	11.8
			(0.912)	0
Sn^{119} in SnTe	0.79_4‖	0.75	0.43 ± 0.05‖	19.4
Sn^{119} in Nb_3Sn	0.84_9¶	0.84	0.44 ± 0.05¶	4

*Values from ref. [5] except where noted.
†All values from Taylor and Craig [2].
‡Values from ref. [14] except where noted.
§D. L. Sprague, cited in ref. [14].
‖From ref. [15].
¶From ref. [16].

of a constant shift [Eq. (4)] to make $(\delta v/c)_T$ approach zero at $T = 0$ and upon the shape of the curve through the data. Even the most precise data available are inadequate below about 20°K. Therefore, we have chosen the constant shift to be that which gives the best fit at low T to $(\delta v/c)_T = AT^4$, the functional form given by the Debye model near $T = 0$. At high temperatures the shift must approach $(\delta v/c) = -3kT/2M'c^2$. A smooth curve was drawn through the data in the intermediate range. Numerical integration of the intermediate region plus the analytical extension at the temperature extremes resulted in the $f(0)$ values given in Table II. The relative contributions of the low- and high-temperature integrals were about 10% and 20%, respectively, so most of the contribution to Eq. (12) is indeed governed by the data. Only three of the seven Fe^{57} systems studied [5] are listed; the recent detailed $f(T)$ studies of Nussbaum *et al.* [14] allow a ready comparison in these cases. The good agreement confirms the Kagan theory.

The application of Eq. (12) to SnTe and Nb_3Sn yields $f(0)$ values much higher than are observed. In view of the large anharmonicity [15, 16], the Kagan theory should not be expected to apply. Curiously, $f(0)$ values determined from

$$f(0) = \exp\{(-3E/4M'c^2)[S(\infty)/2]^{\frac{1}{2}}\} \tag{13}$$

a relation from the Debye model, are almost the same as those found here, even though they do not utilize the same data.

Dash *et al.* [9] have investigated the effects of anharmonicity on the f value. At high temperatures they show that the observed f may be represented by a product $f_h f_a$, where the subscripts refer to a harmonic term and an anharmonic term. Taking f_a to be temperature-independent, Shier and Taylor [16] were able to represent their f data; the harmonic term corresponded to a θ_D of several hundred degrees, again in sharp contrast to that found from $f(0)$ alone and in crude agreement with the ψ value of Table I.

NEW EXPERIMENTAL RESULTS: Fe IN Mo

Better data should allow a more stringent experimental test of the correlation between $\ln f$ and the shift. Improved Mössbauer drives, better utilization of computer-fitting programs for the spectra, better detectors, and improved furnace-cryostat design are now available, and have been employed to obtain the data shown in Fig. 3. An Fe^{57}-in-Mo source was analyzed by a narrow-line, enriched potassium ferrocyanide absorber for shifts and relative f values over the range $4 < T < 750°K$ and by a broad, black lithium ammonium ferrous fluoride absorber for absolute f value calibration. A molybdenum source was chosen in order to allow comparison of $\langle\omega\rangle/\langle\omega^{-1}\rangle$ values calculated by McMillan [1] from non-Mössbauer data. The parameters determined are given in Table I. McMillan's value, a quantity equivalent to ψ_0, is 310°K for Mo. The agreement is poorer if the observed $\psi_0 = 271°K$ for Fe in Mo is modified by the factor $(M_{Fe}/M_{Mo})^{\frac{1}{2}}$ under the assumption that the Fe behaves like a Mo atom except for the mass change.

In Fig. 3 we see that the linearity proposed in ref. [2] is well borne out for $T \gtrsim 0.3\theta_D$ and that the low-temperature turn-up predicted, for example, by the Debye model is observed. The f values determined by the area method using the ferrocyanide absorber have been normalized to the absolute f values determined below room temperature using a calibrated black absorber, resulting in $f_0 = 0.906 \pm 0.02$; the uncertainty will be reduced whenever final analysis of all the data is complete. This $f(0)$ implies $\theta_D = 347°K$. A plot of $\ln f$ *versus* T shows a high-temperature slope consistent with $\theta_D = 379°K$ and no evidence of significant anharmonicity. A plot of $\Delta E/E$ *versus* T indicates precisely the classical behavior at high temperature given by $\Delta E/E = -3kT/2M'c^2$. This supports Eq. (4), which includes the second-order Doppler shift as the only temperature-dependent term necessary to describe the Fe data. Other temperature-dependent terms reflecting the volume change of the host must be compensated [5]. The low-temperature behavior of the shift using a Debye model gives θ_D (shift) $= 277°K$.

SUMMARY

We have suggested that the zero-point mean-square-velocity of Mössbauer nuclei can be determined from a parametric representation of the logarithm of the recoil-free fraction $\ln f(T)$ *versus* $(\delta v/c)_{2D}$, the second-order Doppler shift. We have shown that the method is applicable to a number of metallic systems where the parametric representation appears to be quite linear. Further, the correlation holds for two very anharmonic systems, SnTe [15] and Nb_3Sn [16]. We have also made an experimental test of Kagan's integral relation between $\ln f(0)$ and $(\delta v/c)_{2D}$ and have found results similar to those from the method presented here. Further, we have suggested from calculations within the mass-impurity formalism of Elliott that the $\ln f(T)$ *versus* $(\delta v/c)_{2D}$ curves are nearly linear for the Debye model.

We have also pointed out that in the harmonic approximation the parametric representation, $\ln f(T)$ *versus* $(\delta v/c)_{2D}$, determines $[\langle\omega\rangle/\langle\omega^{-1}\rangle]_{ZP}$, the phonon parameter important in the theory of strongly coupled superconductors. It also appears that heavy impurities in a light lattice tend to decrease $[\langle\omega\rangle/\langle\omega^{-1}\rangle]_{ZP}$ locally, and thereby may increase the superconducting critical temperature. Realistic model calculations should allow Mössbauer measurements of $\ln f(0)$ *versus* $(\delta v/c)_{2D}$ to be related to $[\langle\omega\rangle/\langle\omega^{-1}\rangle]_{ZP}$ of the host lattice.

ACKNOWLEDGMENTS

We express our gratitude to M. Blume and R. E. Watson for useful discussions.

APPENDIX A

Because $S(T)$ becomes virtually independent of temperature for $T > \theta/\sqrt{R}$, it is possible to determine the high-temperature $S(T) = S(\infty)$ from the slope of a $\ln f(T)$ *versus* $\Delta E/E$ plot. Furthermore, it is sufficient to use relative $f(T)$ measurements.

Suppose the relative $f(T)$ measurement at T_i is $\alpha f(T_i)$, where α is independent of temperature and $f(T_i)$ is the absolute recoil free fraction. Then, using the definition of $S(T)$, Eq. (6),

$$\ln[\alpha f(T_i)] - \ln \alpha = \ln f(T_i) = S(T_i)\left\{\left(\frac{\Delta E}{E}\right)_{T_i} - \left[\left(\frac{\delta v}{c}\right)_{\text{iso}} + \left(\frac{\delta v}{c}\right)_{\text{abs}}\right]\right\} \tag{A1}$$

where α and the term in the square brackets are both unknown but temperature-independent. If measurements are made at two temperatures, $T_i = T_1$ and T_2, both greater than $\theta/\sqrt{R}$, $S(T_1) \approx S(\bar{T}) \approx S(T_2)$. Then

$$\ln[\alpha f(T_1)] - \ln[\alpha f(T_2)] = S(\bar{T})\left[\left(\frac{\Delta E}{E}\right)_{T_1} - \left(\frac{\Delta E}{E}\right)_{T_2}\right] \tag{A2}$$

Thus $S(\bar{T})$ is the slope of the experimental plot in the high-temperature region.

Unfortunately, in the low-temperature region it is not possible to determine $S(T)$ from the slope, even though $S(T)$ becomes temperature-independent. The reason for this difficulty is that $(\Delta E/E)$ is also becoming independent of the temperature because $(\delta v/c)_T \ll (\delta v/c)_{ZP}$.

REFERENCES

1. W. L. McMillan, *Phys. Rev.* **167**:331 (1968).
2. R. D. Taylor and P. P. Craig, *Phys. Rev.* **175**:782 (1968).
3. V. I. Goldanskii and R. H. Herber (eds.), *Chemical Applications of Mössbauer Spectroscopy* (Academic Press, New York, 1968).
4. B. J. Josephson, *Phys. Rev. Letters* **4**:341 (1960).
5. W. A. Steyert and R. D. Taylor, *Phys. Rev.* **134**:A716 (1964).
6. W. M. Visscher, *Phys. Rev.* **129**:28 (1963).
7. A. A. Maradudin and P. A. Flinn, *Phys. Rev.* **126**:2059 (1962); K. N. Pathak and B. Deo, *Physica* **35**:167 (1967); R. M. Housley and F. Hess, *Phys. Rev.* **146**:517 (1966).
8. R. J. Elliott, in: *Phonons in Perfect Lattices and Lattices with Point Imperfections*, R. W. H. Stevenson, ed. (Oliver and Boyd, Edinburgh and London, 1966), Chap. 14.
9. J. G. Dash, D. P. Johnson, and W. M. Visscher, *Phys. Rev.* **168**:1087 (1968).
10. Y. Hazony, *J. Chem. Phys.* **45**:2664 (1966).

11. M. G. Clark, G. M. Bancroft, and A. J. Stone, *J. Chem. Phys.* **47**:4250 (1967).
12. J. S. Shier and R. D. Taylor, *Solid State Commun.* **5**:147 (1967).
13. Yu. Kagan, *Zh. Eksperim. i Teor. Fiz.* **47**:366 (1964); *Soviet Phys.—JETP* **20**:243 (1965).
14. R. H. Nussbaum, D. G. Howard, W. L. Nees, and C. F. Steen, *Phys. Rev.* **173**:653 (1968).
15. R. D. Taylor and J. S. Shier (unpublished).
16. J. S. Shier and R. D. Taylor, *Phys. Rev.* **174**:346 (1968).
17. V. J. Minkiewicz, G. Shirane, and R. Nathans, *Phys. Rev.* **162**:528 (1967).
18. D. Bijl, in: *Progress in Low Temperature Physics*, Vol. 2, C. J. Gorter, ed. (North-Holland Publishing Co., Amsterdam, 1957), Chap. XIII.
19. L. J. Vieland and A. W. Wicklund, *Phys. Rev.* **166**:424 (1968).

The Mössbauer Effect in Microcrystals

D. Schroeer

Department of Physics
University of North Carolina
at Chapel Hill, North Carolina

Major modifications of the crystal lattice occur in microcrystals. This paper presents an examination of the effects of these lattice modifications on various aspects of Mössbauer spectra, together with a summary of the relevant literature. Examples discussed include the following: Surface vibrational modes and partial decoupling from the supporting medium can modify the Debye–Waller factor. The Debye temperature is affected by the "internal pressures" (which can be as much as ±200 kbar) *corresponding to the changes in the lattice spacing created by surface effects. These internal pressures can result in isomer shifts. Various chemical valence states may be preferentially stabilized in microcrystals. The finite crystallite size leads to superparamagnetism. Transition temperatures, such as those of the Néel and Morin transitions, may be shifted through the internal pressures. Surface and core distortions through unsaturated chemical bonds and internal pressures can lead to changes in the electric-quadrupole interaction.*

INTRODUCTION

Dr. L. N. Mulay brightened the Third Mössbauer Symposium [1] by his example of a frog as a possible subject for a "let-us-see-what-happens" experiment. The study of microcrystals by means of the Mössbauer effect is a little bit like this frog investigation, in the sense that one frequently is not sure just what to expect, although confident that something will happen. Since microcrystals may occur unexpectedly in samples, and must then be recognized and taken into account, such studies are important for their own sake. But there is more to it than that, as this paper will attempt to show.

Microcrystals are crystallites containing from a few hundred to a few hundred million atoms. If they are not spherical, the words "particle size" or "diameter" will be used to mean $d = (6V/\pi)^{1/3}$, where

V is the microcrystal volume; *d* ranges from a few thousand angstroms down. One outstanding characteristic of microcrystals is that a significant fraction of their atoms lie on or near the surface. This surface not only may itself be considered a huge defect, but it also introduces further defects into the interior of the microcrystal, modifying the whole lattice structure. The other characteristic is that a relatively small number of atoms is contained in any of these microcrystals. This not only limits cooperative effects, but also changes continuous energy-level distributions to discrete sets. The study of microcrystals can give information about these defects, about the nature of cooperative effects, and about all aspects of the crystalline state.

The Mössbauer effect is particularly suitable for microcrystal studies because it directly samples the microscopic surroundings of the emitting or absorbing nucleus. This is in contrast to other techniques which are dependent on gross effects, such as X-ray scattering, with its broadening in microcrystals, or resistivity measurements, which are hampered by the many interfaces in microcrystal samples.

Considerable theoretical and experimental effort has recently been invested in the Mössbauer effect in microcrystals. It therefore seems appropriate to carry out at this time a review (by no means comprehensive) of this field. This review will indicate the major effects which have so far been observed, and will try to provide a basis for future investigations. The review will begin by outlining the procedures for producing microcrystals, and the attendant problems. Next it will examine the resulting lattice modifications which can occur in them. Then it will discuss the changes in microcrystals, relative to bulk, which have been detected in various aspects of Mössbauer spectroscopy. At the end a few conclusions and projections will be presented.

PRODUCTION OF MICROCRYSTALS

Precipitation

Microcrystals of various sizes can be produced by drying and heating a precipitated sample. In fact, microcrystals may sometimes be accidentally produced when low sample-preparation temperatures are used.

For example, samples of α-Fe_2O_3 (hematite) of various average particle sizes have been prepared through precipitation by Takada and Kawai [2] and Yamamoto [3]. The intermediate product is probably α-FeOOH [2], which is then calcined at temperatures between 250°C and 800°C for times on the order of a few hours. At the lower temperatures there may still be residual hydroxide and water molecules attached to the surface of the microcrystals, even after extended calcining. These contaminations may distort the surface and add impurities to the sample.

Alternatively, the intermediate product α-FeOOH may be produced by heating aqueous solutions of iron nitrate, for example [4]. Typical particle sizes in such samples, as obtained by us from the broadening of the X-ray diffraction lines [5] and electron microscopy, are listed in Table I. However, these particle sizes were not very reproducible.

Table I. Average Particle Size in α-Fe_2O_3 Samples Prepared by Heating a Solution of Iron Nitrate for 2 hr

Temperature, °C	Average particle diameter, Å
250	50
300	525
400	1500

The particle-size distribution resulting from precipitation is usually quite broad. Hasegawa and Sato [7] have studied this distribution for $CoFe_2O_4$ microcrystals obtained by coprecipitation [6]. For a sample prepared at 100°C, for example, the average particle size was about 180 Å, with a full-width at half-maximum of about 120 Å.

Hydrosols

Metal microcrystals can be produced in a controlled manner by preparing a hydrosol—a colloidal dispersion of the microcrystals in water. Wilenzick *et al.* [8], for example, prepared microcrystal hydrosols of platinum and gold by the condensation process. The resultant particle size is both reproducible and uniform to better than 20%. However, this method is somewhat limited in the number of different materials which can be produced.

Evaporation

Microcrystals of a wide variety of metals [9–11] and simple compounds [12, 13] may be produced by evaporation techniques.

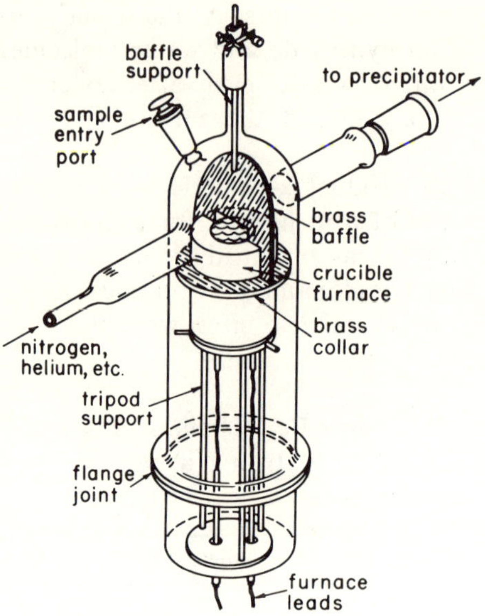

Fig. 1. Apparatus for producing microcrystals *via* evaporation. (After Young and Morrison [12].)

Figure 1 shows a typical apparatus for mass-producing microcrystals by evaporation, as described by Young and Morrison [12]. The bulk sample is heated to liquefaction in vacuum. A flow of an inert gas is passed over it, taking some of the evaporated sample with it. The gas is then fed through a precipitator to remove the microcrystals. There are many variations possible in this procedure. Karioris *et al.* [14], for example, produced gold microcrystals on the order of 400 Å in diameter by vaporizing gold wires by the exploding-wire phenomenon. Suzdalev *et al.* [15] passed vaporized tin through a filter to produce a wide range of tin microcrystals. Buckel *et al.* [10, 11] produced microcrystals of many metals by forming thin films on cryogenically cooled substrates. The particle size in all cases is a function of the particular inert gas used, of the gas pressure and flow rate, etc. For the cooled substrate it is also a function of the substrate temperature. The evaporation technique can be used to produce large quantities of microcrystals, without, however, too much uniformity or reproducibility.

High-Area Silicates

Perhaps the easiest technique for producing large samples of microcrystals of controllable size is that of preparing them on a high-

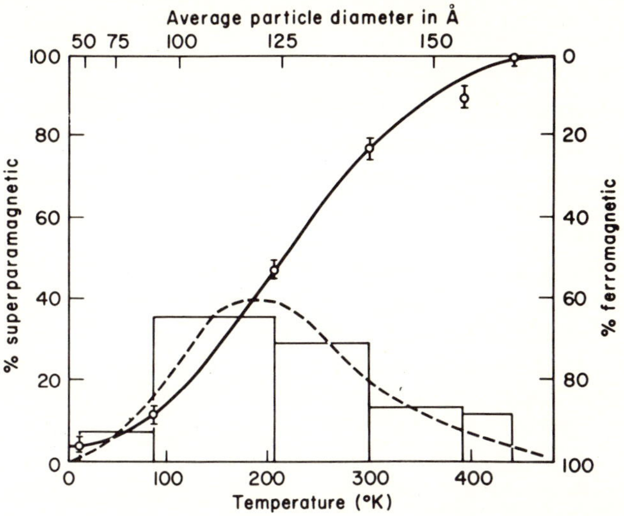

Fig. 2. Particle-size distribution (dashed line) in an 135-Å (avg.) diameter α-Fe_2O_3 microcrystal sample produced by calcining silica gel impregnated with iron nitrate. The distribution was derived from the ratio of single-line to hfs in the Mössbauer spectra. (After Kündig *et al.* [16].)

area silicate. Microcrystals of α-Fe_2O_3 [16], NiO [17], and Ni [18], for example, have been made by impregnating silica gel with a nitrate solution and then calcining the sample in dry air [16, 17] or in a hydrogen atmosphere [18]. A typical particle-size distribution is shown in Fig. 2 (see below for comments concerning how to determine such a size distribution). Advantages of this preparative technique are that large quantities can be readily prepared, that samples can be stored in air without oxidizing, that the average particle size can be easily varied by changing the concentration of the impregnating liquid, and that the sample temperature can be raised as high as 500°C without fusing the microcrystals together (since they stay inside the pores of the silica gel). There is the drawback that there may be unknown chemical formations occurring at the interface between the microcrystals and the silica gel.

Miscellaneous

Other possible techniques include the production of clusters inside host materials [19–23], forcing soft metals under pressure into porous glass [24], and burning Mg to obtain MgO microcrystals [25].

LATTICE MODIFICATIONS IN MICROCRYSTALS

Surface and Volume Effects

The surface *per se* in microcrystals permits surface modes of vibration, which make contributions to the phonon spectrum [26, 27]. The volume limitation introduces a long-wavelength cutoff to the phonon spectrum [26, 27]. The volume also limits the size of the domain which can be involved in any cooperative or long-range order phenomena, such as ferro- and anti-ferromagnetism, leading to the often-observed phenomenon of superparamagnetism. The small number of atoms which are included in these microcrystals changes continuous energy-level distributions, such as those of electrons in metals, into discrete sets. If the separation of these energy levels becomes comparable to the thermal energy kT, changes may occur in the gross properties of the crystal lattice [28].

Stresses and Strains

The existence of a surface means that there will be atoms in an asymmetrical environment whose binding to the remaining crystal will no longer be the same as in bulk [28]. There will be a surface free-energy which can be extrapolated from the liquid-state surface tension [14, 29]. Lennard-Jones [30] pointed out some time ago that if one assumes some realistic crystalline potentials, then the average lattice spacing for ionic crystallites should be contracted, and that for covalent crystallites should be expanded. More recently a discussion concerning these points has been going on in [31–33]. Lastly, the surface layers may be very strongly distorted [34, 35], which may result in anisotropic deformations of the whole microcrystal volume.

These surface effects will result in lattice distortions which should be inversely proportional to the particle diameter, similar to the pressure from surface tension in liquid drops. They have been frequently observed experimentally. Figure 3 shows the Co K_α X rays scattered from the (104) plane of α-Fe_2O_3 microcrystals of various sizes. The broadening in the peak is due to the particles size, the shift in peak position is due to a change in the lattice spacing. The data gathered from such X-ray scattering peaks gives Fig. 4, where the lattice spacing in these microcrystals is plotted as a function of 1/(particle size) [4]. Figure 5 shows similar data for condensed gold vapor [14]. As predicted by Lennard-Jones [30], Boswell finds that the lattice in ionic alkali halide microcrystals contracts linearly with

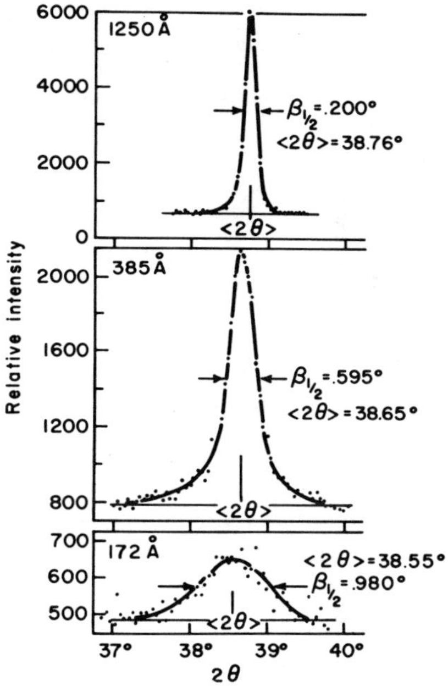

Fig. 3. Co K_α X rays scattered from the (104) planes of α-Fe_2O_3 microcrystals of various size. The broadening is due to particle size, the shift is due to a change in the lattice spacing.

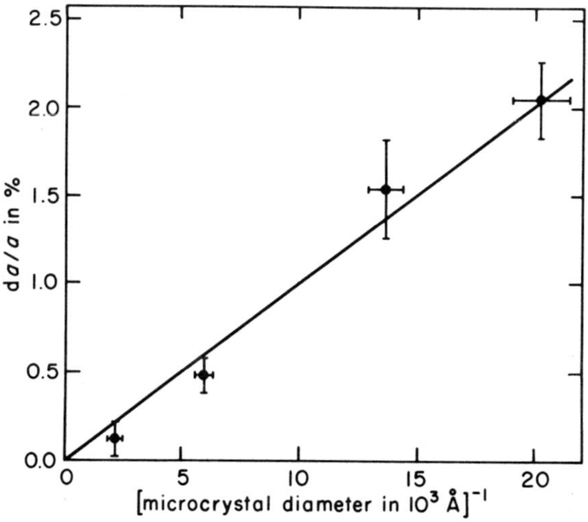

Fig. 4. The lattice spacing as a function of 1/(particle diameter) for α-Fe_2O_3 microcrystals. (After Schroeer and Nininger [4].)

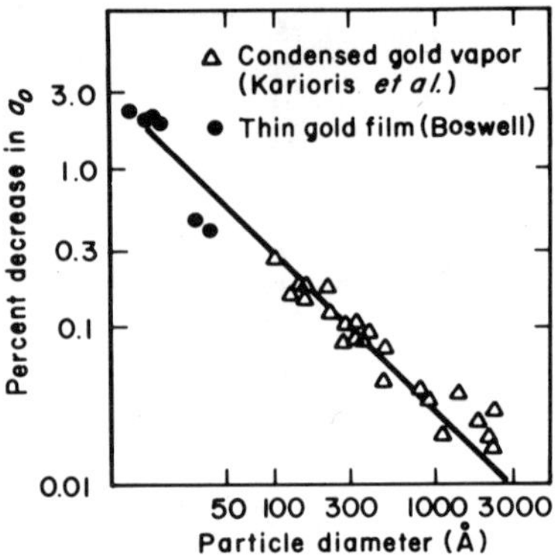

Fig. 5. The lattice spacing as a function of particle size for gold microcrystals produced by evaporation onto a substrate. (After Karioris *et al.* [14].)

1/(particle diameter), as if they were under positive external pressure. One can consider a separate divalent molecule to be the extreme example of a microcrystal, and the interatomic separations in the alkali halides agree roughly with an extrapolation of the microcrystal data. In the more covalent α-Fe_2O_3 microcrystals the lattice expands as if it were under external negative pressure (as high as -200 kbar). In this paper such lattice changes will be called *internal pressures*; they can perhaps be considered a poor man's high-pressure apparatus.

Defects

The surface of a microcrystal may be thought of as a huge defect, leading to some of the stresses and strains discussed above. As a result, the surface may be in a different chemical state than the interior. For example, there may be gas adsorbed, such as oxygen from the air. In metallic microcrystals this can lead to a fairly thick layer of oxide on the surface. When going to α-Fe_2O_3 microcrystals *via* the intermediate product of α-FeOOH there may be some moisture retained even at elevated temperatures, leading to a thin layer at the surface which is different from the core. Similar effects occur in the production of microcrystals of MgO from brucite ($MgO \cdot H_2O$) [31]. These layers can lead to differences in chemical bonding in the outer lattice layers, and to distortions which may spread through the whole microcrystal volume [36] and result in internal pressures.

The surface may also increase the number of defects inside the microcrystals. Transition-element oxides are notoriously nonstoichiometric; bulk CoO, for example, can have oxygen excesses as large as 1.2% [37]. When these oxides are prepared as microcrystals the increased surface may be expected to enhance this nonstoichiometry. Table II lists the oxygen excesses observed in NiO microcrystals [38]. These surface-induced defects can lead to changes in the

Table II. Oxygen Excess in NiO Microcrystals*

Preparation temperature, °C	250	300	350	400	500	600	700	1300
Microcrystal size, Å	80	104	117	140	170	664	>1000	>2000
Oxygen excess, %	0.36	0.37	0.32	0.33	0.15	0.11	0.03	0.00

*After Richardson and Milligan [38].

microcrystal lattice spacing [36], and hence can make contributions to the internal pressure. The term "controlled valency materials" has been suggested for these nonstoichiometric oxides [37], and possible effects on the chemical valence of the component ions might be expected.

Coupling to the Environment

These microcrystals differ from bulk in that a significant fraction of their atoms are involved in the coupling to the surroundings, which may be quite different in structure. This coupling may have chemical effects at the interface, it may introduce distortions and defects, it may influence long-range order, and it can modify the phonon spectrum.

MODIFICATIONS OF THE MÖSSBAUER EFFECT IN MICROCRYSTALS

The lattice modifications that have just been discussed will modify the Mössbauer spectra of these microcrystals. Table III summarizes both observed and expected effects in the Debye–Waller f factor, the isomer shift, the magnetic interaction, and the quadrupole interaction. Brief discussions of some of these observed effects will now be presented.

Debye–Waller f Factor

Changes in the Mössbauer recoil-free fraction f in microcrystals have been observed by several investigators [15, 26, 39–42]. It has been quantitatively investigated by Marshall and Wilenzick [26]. They

Table III. Lattice Modifications, and Resulting Changes in Mössbauer Spectra (Italicized), Which Do or May Occur in Microcrystals

	f factor	Isomer shift	Magnetic interaction	Quadrupole interaction
Surface	Surface phonon modes *changes in $f(T)$* [26]	Surface phonon modes *small 2nd-order Doppler shift*	—	—
Volume	Long-wavelength phonon cutoff *changes in $f(T)$* [26] Energy-level quantization *changes in $f(T)$*	Long-wavelength cutoff *small 2nd-order Doppler shift* Superparamagnetism *transition isomer shift* [16, 46] Energy-level quantization *isomer shifts*	Superparamagnetism *collapse of hfs* [16–18, 52, 53] *restoration of hfs with applied magnetic field* [18, 55] Change in long-range order and cooperative effects *change in H, T_c, T_N, etc.* [4, 47, 48] Annealing *restoration of hfs* [47, 48] Energy-level quantization *changes in hfs*	Change in Sternheimer factors *change in quadrupole interaction* Relaxation effects *unequal peak intensities* Energy-level quantization *change in quadrupole interaction*

Stress and strains	Internal pressure *changes in $f(T)$ through* θ_D[15, 26, 39–42] Surface anisotropy $\theta_{\perp} \neq \theta_{\parallel}$ [35]	Internal pressure *2nd-order Doppler shift, isomer shift* [51] Surface anisotropy *anisotropy in Doppler shift* Surface bonding *isomer shifts*	Internal pressure *change in T_c, T_M, etc.* [4, 47, 48] Net magnetic moment *antiferro to weak ferromagnetism* Phase changes *changes in H, etc.*	Distortion at surface *V_{zz} larger* [61–63] *line broadening* Distortion of whole lattice *change in V_{zz}* [61–63] Phase change *change in V_{zz}*
Defects	Internal pressure *changes $f(T)$* [15, 26, 39–42] Nearby vacancies *changes in $f(T)$*	Introduction of oxygen and defects *valence change* [17, 47, 48] Internal pressure *isomer shift* [51]	Net magnetic moment *applied magnetic field restores hfs* [18, 55] Internal pressure *changes in T_c, T_M, etc.* [4, 47, 48] Phase changes *change in H, etc.*	Internal pressure *change in V_{zz}* [61–63] Phase change *change in V_{zz}* Distortion *line broadening*
Coupling	Possible recoil *reduces $f(T)$, makes $f(T)$ support-dependent* [41] Phonon node at surface *$f(T)$ support-dependent* [26]	Chemical bonding to surroundings *isomer shift*	Coupling to magnetic field of surroundings *change in H, polarization*	Distortions at interface *changes in V_{zz}, line broadening*

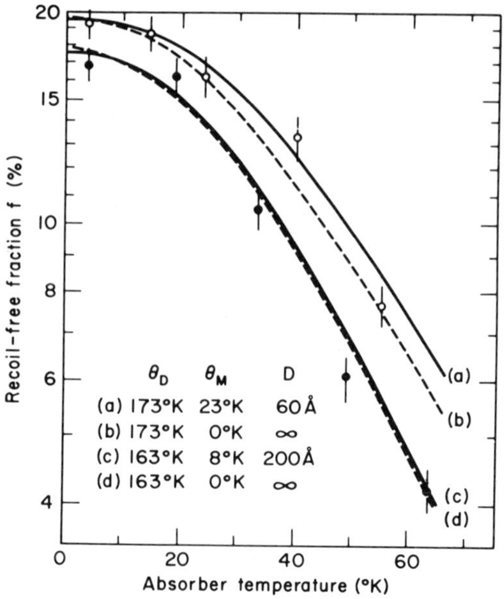

Fig. 6. The recoil-free fraction f in gold microcrystals 200 Å (darkened circles) and 60 Å (open circles) in diameter. The dashed curves are theoretical curves not taking the surface into account, while the solid curves are least-squares fits to the data and do take the surface into account. (After Marshall and Wilenzick [26].)

studied the recoil-free fraction *versus* the temperature in gold microcrystals of 200 Å and 60 Å diameter embedded in gelatin. Figure 6 shows the data obtained. In analyzing it, Marshall and Wilenzick took into account the expected surface phonon modes, the long-wavelength phonon cutoff, and a possible change in the Debye temperature θ_D (the low-wavelength cutoff). The surface waves are introduced by replacing the Debye phonon distribution by the approximation

$$z(\omega) \propto \{(4\pi V/c)\omega^2 + (3\pi S/4)\omega\}$$

where ω is the phonon frequency, V and S are, respectively, the volume and the surface of the microcrystal, and c is the phonon propagation velocity. The long-wavelength cutoff is introduced through the lower cutoff temperature $\theta_M = h\omega_{\min}/k$. There are then two constants to be fitted by the data, namely, θ_M and θ_D. A more-detailed theoretical expression with additional parameters has been derived by Thiel [27].

The dashed curves in Fig. 6 show the behavior of $f(T)$ with no surface effects taken into account; the solid curves are the least-squares fits to the data with the values of θ_M and θ_D shown in Table IV.

Table IV. Experimental and Theoretical Values of the Debye Temperature θ_D and Lower Cutoff Temperature θ_M for Gold Microcrystals*

Particle size, Å	θ_M, °K Experimental	Rigid surface	Free surface	θ_D, °K Experimental	Theoretical
200	8 ± 2	2.3	8.6	163	Assume 163
60	23 ± 6	7.8	28.7	173	171

*Marshall and Wilenzick [26].

The different theoretical values of θ_M were obtained assuming either a free or a rigidly bound microcrystal surface; the values for a free surface clearly agree much better with the experimental data. This means that the binding of these microcrystals to the gelatin medium was relatively loose, as one would expect, since the acoustical impedance mismatch between gold and gelatin is quite large. The effect of the surface vibrational modes and of θ_M is to increase the f factor, with a larger increase at higher temperatures. The rise in the Debye temperature has been related [40] to the change in lattice spacing observed in gold microcrystals [13, 14]. The Debye temperature of $\theta_D = 171$°K for the 60-Å sample is obtained by taking $\theta_D = 163$°K for the 200-Å sample with

$$\Delta\theta = \theta[1 - \gamma(\Delta V/V)]$$

where γ is the Grüneisen constant (~ 3), V is the volume of the elementary unit cell in the 200-Å microcrystals, and $\Delta V = V(60\text{ Å}) - V(200\text{ Å})$ [43].

Suzdalev *et al.* [15] found the Mössbauer Debye temperature in tin microcrystals to be particle-size dependent, as shown in Fig. 7. The reasonably straight-line fit indicates the same 1/(particle diameter) dependence as in all surface-dependent effects. Suzdalev *et al.* interpret this effect as being due to a combination of a central core with a $\theta_D = 100$°K and five surface layers with $\theta_D = 140$°K. Such changes in θ_D in microcrystals have also been observed in Mössbauer studies of tungsten [39] and in several non-Mössbauer studies [10, 42, 45].

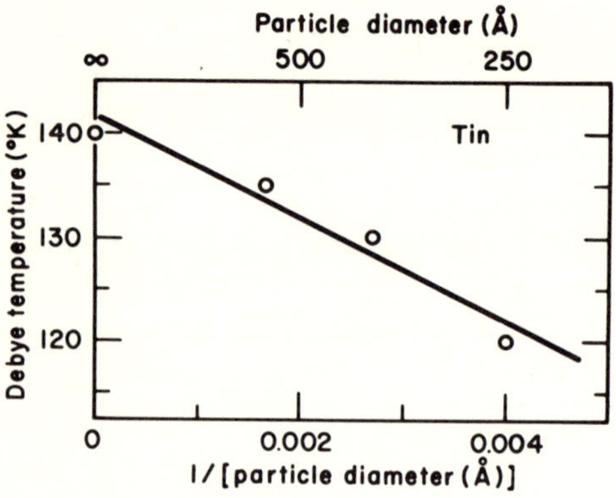

Fig. 7. The Mössbauer Debye temperature in tin microcrystals as a function of 1/(particle diameter). (After Suzdalev *et al.* [15].)

The effect of the coupling to the environment on the ability of the microcrystals to sustain a Mössbauer effect has been investigated by Van Wieringen [41], who notes that a mass of at least 4×10^5 iron atoms must absorb the recoil momentum of the nuclear decay to avoid a significant line broadening and hence a reduction in the Mössbauer effect. He observed an increase in the f factor of iron oxide microcrystals when they are glued to a substrate rather than just being loosely pressed together (see also [42]), and ascribes this increase to the improved coupling to the surroundings of the glued powder.

Isomer Shift

A few cases of isomer-shift differences between bulk and microcrystals have been reported [16, 17, 46–48]. One would expect all modifications in the phonon spectrum to result in apparent isomer shifts through the second-order Doppler shift. Most of these effects will be small, but the change in θ_D may lead to a detectable isomer shift. For example, in an iron-containing compound with a bulk Debye temperature of 300°K a 10% increase in θ_D (corresponding to a ~1% decrease in the lattice spacing) would lead to an apparent isomer shift of about 0.01 mm/sec at room temperature.

The internal pressure in microcrystals may also result in a change in the electron density and hence in the isomer shift. In gold microcrystals of 200-Å and 60-Å diameter the internal pressures are 12 kbar and 40 kbar, respectively (from Fig. 5). Patterson *et al.* [49, 50] have studied the isomer shift in bulk gold under pressure. From their data one would predict isomer shifts of about +0.05 and +0.14 mm/sec, respectively, in the 200-Å and 60-Å microcrystals. Preliminary experiments for a 200-Å sample have given an isomer shift of $+0.034 \pm 0.012$ mm/sec [51], in satisfactory agreement with the prediction.

A negative isomer shift in α-Fe_2O_3 microcrystals as compared to the bulk value has been observed [16, 46]. These shifts are on the order of 0.05 mm/sec. Their exact dependence on particle size and their origins are not clear. The isomer shift in bulk is not very pressure-sensitive [46], so that the internal negative pressure should contribute little to these shifts. However, a part of these shifts may be related to the collapse of the magnetic hyperfine structure (hfs) at the superparamagnetic transition (see below for discussion of superparamagnetism).

A particularly interesting class of isomer-shift differences has been observed by Ando *et al.* [17] in NiO microcrystals doped with Co^{57}. Figures 8 and 9 show Mössbauer spectra of NiO : Co^{57} as bulk and as 103-Å particles, respectively. In the bulk sample both Fe^{2+} and Fe^{3+} valence states occur; in the microcrystal sample only the Fe^{3+} valence state occurs. Ando *et al.* [17] associate the enhancement of the Fe^{3+} state in the NiO microcrystals with their increased oxygen excesses [38]. Similar effects have been observed in what appears to be microcrystals of CoO [47, 48*].

Magnetic Interaction

One of the most striking phenomena observed in microcrystals by means of the Mössbauer effect is superparamagnetism [16, 52, 53]. Superparamagnetism has been discussed in detail elsewhere [23], so I shall present here only one example of the possible effects observed.

Each microcrystal of a ferro- or antiferromagnetic material constitutes a single domain, in which the hyperfine field fluctuates randomly with a characteristic relaxation time $\tau = \tau_0 \exp\{+KV/kT\}$, where $1/\tau_0$ is the relaxation frequency in the absence of any anisotropy, K is the anisotropy constant (energy per unit volume), V is the volume of the microcrystal, and kT is the thermal energy [16]. This means that KV is the energy barrier which the magnetization must overcome in going from one easy direction to another. Lindquist *et al.* [18] studied the Mössbauer spectra of 60-Å Ni microcrystals as a function of the temperature and applied external magnetic field. Figure 10 shows such spectra. If $kT \ll KV$, then there is a magnetic hfs in the Mössbauer spectrum; if $kT \gg KV$, then this has collapsed to a single line.

*Electron microscopy has shown that the CoO(II) samples discussed in [48] are microcrystals.

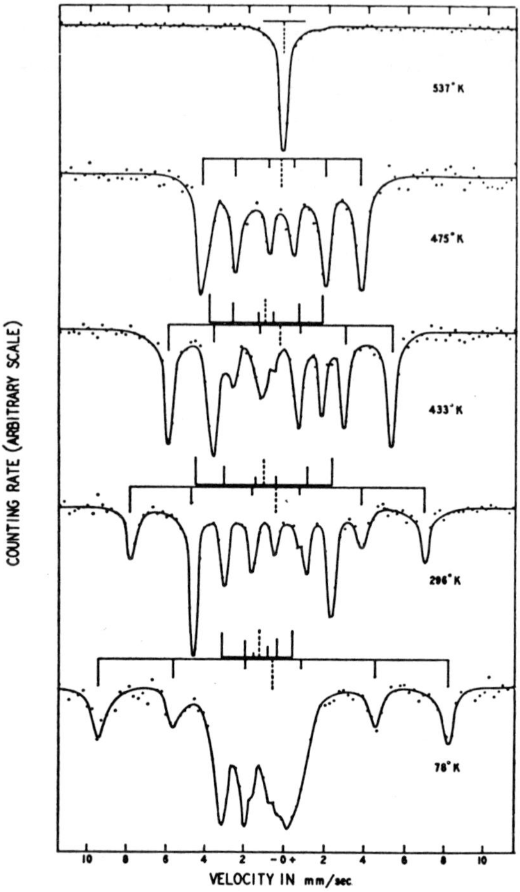

Fig. 8. Mössbauer spectra of bulk NiO doped with Co^{57}, taken at several temperatures. The spectra are analyzed in terms of Fe^{2+} (upper) and Fe^{3+} (lower) valence states. (After Ando *et al.* [17].)

For $kT \approx KV$ we see a partially relaxed magnetic hfs, complicated by the fact that there is a broad distribution in the sizes of the particles. The relaxation temperature yields the value of the anisotropy energy density K. From the particular temperature dependence of the ratio of single-line to hfs in the spectra one can extract the particle size distribution [16], as in Fig. 2, and from the spectrum shape one can get information about the relaxation time involved [18].

If an external magnetic field is applied to the microcrystal sample, one would expect the fluctuations to slow down. When the magnetic

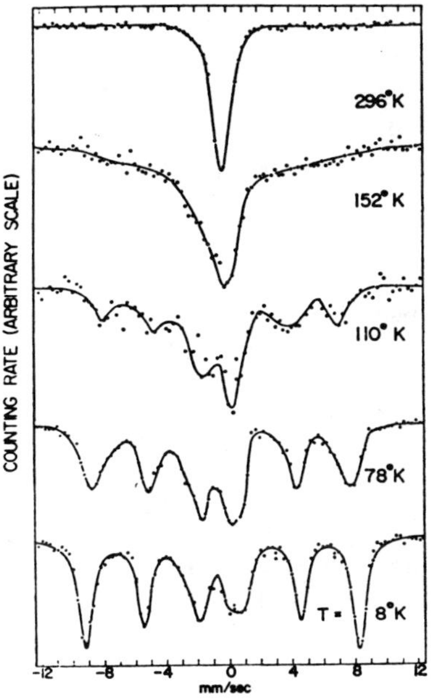

Fig. 9. Mössbauer spectra of 103-Å-diameter NiO microcrystals doped with Co^{57}, at several temperatures. The spectra show only the characteristics of the Fe^{3+} valence state. (After Ando *et al.* [17].)

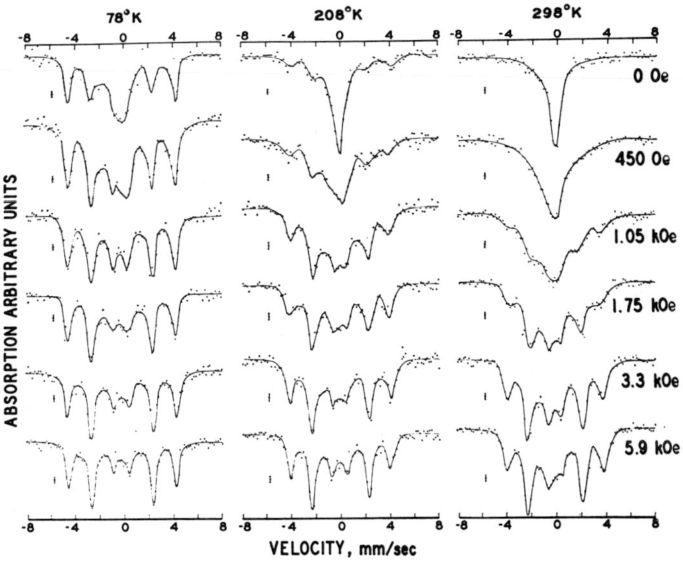

Fig. 10. Mössbauer spectra of 60-Å-diameter Ni microcrystals doped with Co^{57} The spectra were taken at various temperatures and with different externally applied magnetic fields. (After Lindquist *et al.* [18].)

interaction energy is equal to the thermal energy $\mu H = kT$, then the magnetic hfs should be restored (here μ is the magnetic moment of the microcrystal). In Fig. 10 for the Ni particle sample this occurs at 298°K and about 10^3 G. Solving for the effective magnetic moment per Ni atom, we obtain the value

$$\mu_{\text{eff}} = \frac{\mu}{N} = \frac{kT}{HN} = \frac{kTm}{H\rho V} \approx 0.5\mu_B$$

where N is the number of Ni atoms in the microcrystal, ρ the density, V the volume of the microcrystal, and m the mass of an Ni atom. This value is in agreement with the value of $0.6\mu_B$ as reported by Kittel [54].

Similar experiments involving the application of external magnetic fields to superparamagnetic microcrystals have been reported by Eibshütz and Shtrikman [55]. In their analysis they take into account a possible polarization of the microcrystals. Even in antiferromagnetic microcrystals one might expect external magnetic fields to have some effects, because they usually exhibit a small net magnetic moment due to some uncompensated magnetic spins at the surface [56]. However, because these net magnetic moments are small, very large external fields or very low temperatures would be required.

Since ferro- and antiferromagnetic phenomena are cooperative effects, one might expect the magnetic fields and the magnetic transition temperatures in microcrystals to be slightly different, simply because the number of atoms is limited. An even more prominent effect, however, should be the influence of the internal pressure on the magnetic transition temperatures in these microcrystals. Such an effect has been observed in the Morin transition in α-Fe_2O_3 [3, 4]. The Morin transition temperature in bulk rises by 3.6°K/kbar under external pressure [57–59]. Negative external pressure has been applied to single crystals of α-Fe_2O_3 by a "stretching" device, and the Morin transition temperatures goes down at the appropriate rate [36]. The Mössbauer investigations of the Morin transition in α-Fe_2O_3 microcrystals indicate that it is depressed under the negative internal pressure at a rate comparable with the increase observed in bulk under externally applied pressure [4]. Figure 11 summarizes these experimental data, which are in agreement with theory [60]. The Morin transition in the microcrystals is fairly sharp; the spread in it is consistent with typical particle-size distributions [4]. This supports the viewpoint that the lattice changes in microcrystals are homogeneous

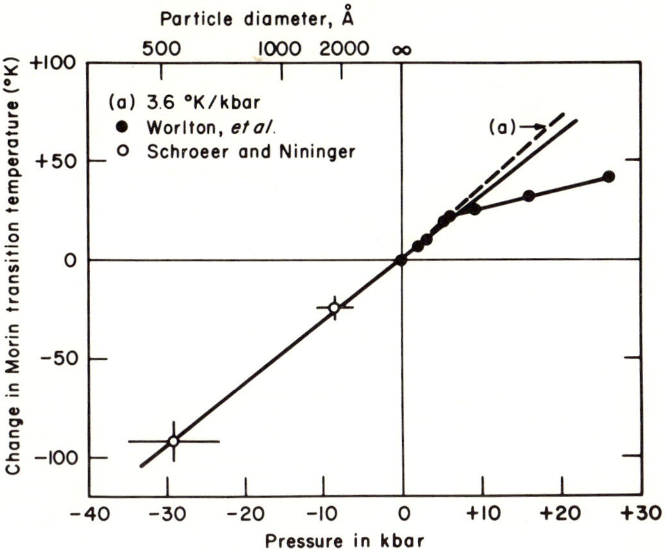

Fig. 11. The Morin transition in α-Fe_2O_3 as a function of pressure. The negative pressures are internal to microcrystals. (After Worlton *et al.* [59] and Schroeer and Nininger [4].)

throughout the whole microcrystal volume; if the microcrystal surface were extremely distorted while the core had bulk condition [16, 61], then the Morin transition would spread over a much wider range of temperatures.

Quadrupole Interaction

With all of the lattice modifications in microcrystals, one might expect that the quadrupole interaction would be different from that in bulk. Such differences have been observed in α-Fe_2O_3 for example [16, 61, 62]. Figure 12 shows the quadrupole splitting in microcrystals of this material as a function of 1/(particle size), as reported by Kündig *et al.* [61]. Hobson and Campbell [62] obtained the very large splitting of 1.87 mm/sec for an indeterminate, though very small, particle size. This 1/(particle size) dependence indicates that this is a surface effect. Kündig *et al.* [61] analyzed the data in terms of a bulk core with a quadrupole splitting of 0.79 mm/sec and a 3.5-Å surface layer with a quadrupole splitting of 1.38 mm/sec. Schroeer [63] found the microcrystal quadrupole-splitting variation to be comparable with that occurring under thermal expansion and notes that the electric field gradient is very sensitive to lattice distortions [46]. He concludes

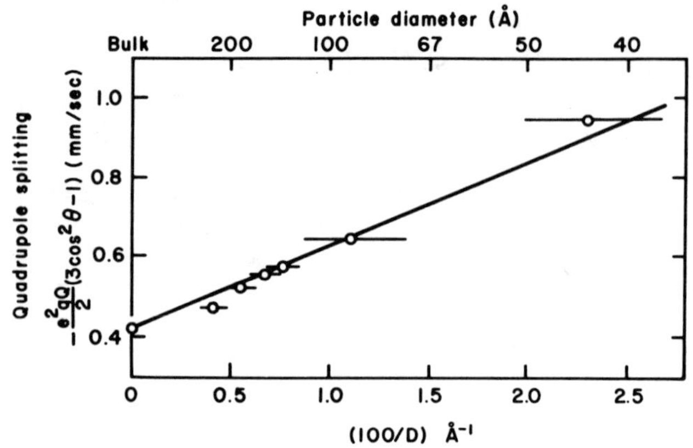

Fig. 12. The quadrupole splitting in α-Fe_2O_3 microcrystals as a function of 1/(particle diameter). (After Kündig *et al.* [61].)

that the increased splitting may be due to the lattice expansion in these microcrystals resulting in a displacement of the lattice components inside the lattice unit cell.

CONCLUSIONS

There are many differences in the properties of microcrystals as opposed to bulk. Some of these have been observed through Mössbauer spectroscopy and have raised at least two important questions: (1) Do the major lattice distortions in microcrystals occur only at the surface [15, 28, 34, 61], or are they spread more or less homogeneously throughout the whole microcrystal volume [4, 30, 36, 40, 63]? The answer to this question depends on the nature of crystalline potentials, chemical bonds, and surface defects. (2) Are the valence differences observed following K-capture decays in microcrystals of nonstoichiometric oxides due to surface effects, and, if so, what are their origins [17, 47, 48]? The explanation of this phenomenon promises to contribute to our understanding of Auger aftereffects, of defects, of valence stabilization, of diffusion processes, etc.

Future investigations into microcrystal properties should of course include general exploratory studies. Not only would such studies add to our basic understanding of the crystalline state, but they would also be a service, since microcrystals find wide industrial applications. Table III indicates some of the more obvious as yet unexplored directions which such future investigations might take.

To summarize, one might say that the Mössbauer effect in microcrystals presents a wide variety of phenomena, all of them interesting and worthy of study. Many of these phenomena have not yet been investigated, and for many of them ingenuity (and careful microcrystal preparation) can be substituted for enormous and expensive experimental effort and equipment. This means that the Mössbauer effect in microcrystals can still be fun.

ACKNOWLEDGMENTS

The initial readings and thoughts on this topic were carried out and developed during a stay as a NATO postdoctoral fellow in 1965–66 at Prof. R. L. Mössbauer's Institut of the Technische Hochschule München. I would like to take this opportunity to express my appreciation for this hospitality. Since then this work has been supported by the Materials Research Center of the University of North Carolina under contract SD–100 with the Advanced Research Projects Agency. Thanks are due to the New England Nuclear Corporation for making this presentation possible, and to S. W. Marshall for past helpful discussions and comments and for his critical reading of the manuscript.

REFERENCES

1. L. N. Mulay, *Mössbauer Effect Methodology*, Vol. 3, I. J. Gruverman, ed. (Plenum Press, New York, 1967), p. 79.
2. Takada and N. Kawai, *J. Phys. Soc. Japan* **17** (Suppl. B-I):691 (1962).
3. N. Yamamoto, *J. Phys. Soc. Japan* **24**:23 (1968).
4. D. Schroeer and R. C. Nininger, Jr., *Phys. Rev. Letters* **19**:632 (1967); R. C. Nininger, Thesis (Physics), University of North Carolina, Chapel Hill, North Carolina (1967).
5. H. P. Klug and L. E. Alexander, *X-Ray Diffraction Procedures* (John Wiley and Sons, New York, 1954).
6. W. J. Schuele and V. D. Deetscreek, *J. Appl. Phys.* **32**(Suppl.):235S (1961).
7. K. Hasegawa and T. Sato, *J. Appl. Phys.* **38**:4707 (1967).
8. R. M. Wilenzick, D. C. Russell, R. H. Morriss, and S. W. Marshall, *J. Chem. Phys.* **47**:533 (1967).
9. K. Kimoto, Y. Kamiya, N. Nonoyama, and R. Uyeda, *Jap. J. Appl. Phys.* **2**:702 (1963); K. Kimota and I. Nichida, *Jap. J. Appl. Phys.* **6**:1047 (1967).
10. W. Buckel and R. Hilsch, *Z. Physik* **131**:420 (1952); **138**:109 (1954).
11. H. Bülow and W. Buckel, *Z. Physik* **145**:141 (1956).
12. D. M. Young and J. A. Morrison, *J. Sci. Instr.* **31**:90 (1954); F. W. Thompson, G. S. Rose, and J. A. Morrison, *J. Sci. Instr.* **32**:325 (1955).
13. F. W. C. Boswell, *Proc. Phys. Soc.* (*London*) **54A**:465 (1951).
14. F. G. Karioris, J. J. Woyci, and R. R. Buckrey, in: *Advances in X-ray Analysis* Vol. 10 (Plenum Press, New York, 1967), p. 250.
15. I. P. Suzdalev, M. Ya. Gen, V. I. Goldanskii, and E. F. Makarov, *Soviet Phys.—JETP* **24**:79 (1967).
16. W. Kündig, H. Bömmel, G. Constabaris, and R. H. Lindquist, *Phys. Rev.* **142**:327 (1966).
17. K. J. Ando, W. Kündig, G. Constabaris, and R. H. Lindquist, *J. Phys. Chem. Solids* **28**:2291 (1967).

18. R. H. Lindquist, G. Constabaris, W. Kündig, and A. M. Portis, *J. Appl. Phys.* **39**:1001 (1968).
19. V. G. Bhide, H. C. Bhasan, and G. K. Shenoy, *Physics Letters* **24A**:109 (1967).
20. K. E. Easterling and H. M. Miekk-Oja, *Acta Met.* **15**:1133 (1967).
21. D. J. Sinkin, P. J. Ficalora, and R. A. Bernhaim, *Physics Letters* **19**:536 (1965).
22. M. Ron, H. Shechter, and S. Niedzwiedz, *J. Appl. Phys.* **39**:265 (1968).
23. D. W. Collins, J. T. Dehn, and L. N. Mulay, in: *Mössbauer Effect Methodology*, Vol. 3, I. J. Gruverman, ed. (Plenum Press, New York, 1967), p. 103.
24. J. H. P. Watson, *Phys. Rev.* **148**:223 (1966); *Physics Letters* **25A**:326 (1967).
25. E. Kinder. *Naturwiss.*,**31**:149 (1943).
26. S. W. Marshall and R. M. Wilenzick, *Phys. Rev. Letters* **16**:219 (1966).
27. R. C. Thiel, *Z. Physik* **200**:227 (1967).
28. A. Kubo, *J. Phys. Soc. Japan* **17**:975 (1962); in *Jubilee Book in Honor of Professor Kastler*, Société Française de Physique (to be published).
29. J. Vail, *Can. J. Phys.* **45**:2651, 2661 (1967).
30. J. E. Lennard-Jones, *Z. Krist.* **75**:215 (1930).
31. R. C. Garvie, *Mat. Res. Bull.* **1**:161 (1966); **2**:897 (1967).
32. P. J. Anderson and A. Scholz, *Mat. Res. Bull.* **2**:895 (1967); **3**:535 (1968).
33. R. M. Curtis, *Mat. Res. Bull.* **3**:533 (1968).
34. A. A. Maradudin and J. Melngailis, *Phys. Rev.* **133**:A1188 (1964); A. A. Maradudin, in: *Solid State Physics*, Vol. 19, F. Seitz and D. Turnbull, eds. (Academic Press, New York, 1966), p. 1.
35. J. W. Burton and R. P. Godwin, *Phys. Rev.* **158**:218 (1967).
36. N. Kawai, F. Ono, and K. Hirooka, *J. Appl. Phys.* **39**:712 (1968).
37. B. Fisher and D. S. Tannhauser, *J. Chem. Phys.* **44**:1663 (1966).
38. J. T. Richardson and W. O. Milligan, *Phys. Rev.* **102**:1289 (1958).
39. S. Roth and E. M. Hörl, *Physics Letters* **25A**:299 (1967).
40. D. Schroeer, *Physics Letters* **21**:123 (1966).
41. J. W. Van Wieringen, *Physics Letters* **26A**:370 (1968).
42. T. Nakamura and S. Shimizu, *Bull. Inst. Chem. Research, Kyoto* **42**:299 (1964).
43. R. V. Hanks, *Phys. Rev.* **124**:1319 (1961).
44. W. Mönch, *Z. Physik* **170**:93 (1962).
45. K. H. Rieder and E. M. Hörl, *Phys. Rev. Letters* **20**:209 (1968).
46. R. W. Vaughan and H. G. Drickamer, *J. Chem. Phys.* **47**:1530 (1967).
47. H. M. Ok and J. G. Mullen, *Phys. Rev.* **168**:550 (1968); *Phys. Rev. Letters* **21**:823 (1968).
48. D. Schroeer and W. Triftshäuser, *Phys. Rev. Letters* **20**:1242 (1968).
49. D. O. Patterson, PhD Thesis (Physics), University of Tennessee, Memphis, Tennessee (1967).
50. L. D. Roberts, J. O. Thomson, and D. O. Patterson, *Bull. Am. Phys. Soc.* **11**:49 (1966); L. D. Roberts, D. O. Patterson, T. O. Thomson, and R. P. Levy, *Phys. Rev.* **179**:656 (1969).
51. D. J. Erickson, S. W. Marshall, R. M. Wilenzick, and D. Schroeer (to be published).
52. W. J. Schuele, S. Shtrikman, and D. Treves, *J. Appl. Phys.* **36**:1010 (1965).
53. T. Shinjo, *J. Phys. Soc. Japan* **21**:917 (1966).
54. C. Kittel, *Introduction to Solid State Physics*, 2nd ed. (John Wiley and Sons, New York, 1956), p. 407.
55. M. Eibschütz and S. Shtrickman, *J. Appl. Phys.* **39**:997 (1968).
56. I. S. Jacobs and C. P. Bean, in: *Magnetism*, Vol. 3, G. T. Rado and H. Suhl, eds. (Academic Press, New York, 1963), p. 275.
57. R. C. Wayne and D. H. Anderson, *Phys. Rev.* **155**:496 (1967).
58. H. Umebayashi, B. C. Frazer, G. Shirane, and W. B. Daniels, *Physics Letters* **22**:407 (1966).
59. T. G. Worlton, R. B. Bennion, and R. M. Brugger, *Physics Letters* **24A**:653 (1967).
60. C. W. Searle, *Physics Letters* **25A**:256 (1967).
61. W. Kündig, K. J. Ando, R. H. Lindquist, and G. Constabaris, *Czech. J. Phys.* **B17**:467 (1967).
62. M. C. Hobson, Jr., and A. D. Campbell, *J. Catalysis* **8**:294 (1967).
63. D. Schroeer, *Physics Letters* **27A**:507 (1968).

Aftereffects of Auger Ionization Following Electron Capture in Cobalt Complexes*

A. Nath and M. P. Klein

Laboratory of Chemical Biodynamics
Lawrence Radiation Laboratory
University of California
Berkeley, California

and

W. Kündig and D. Lichtenstein

University of California
Los Angeles, California

Several cobalt complexes labeled with Co^{57} *were studied. In complexes, where cobalt is held in a large and highly conjugated ring system, it is shown that the bonds escape fragmentation after the Auger event. On the other hand, in some complexes with nonconjugated structures, there is evidence for complete fragmentation. In the intermediate category of complexes, there exists a certain probability for the molecule to escape fragmentation. Attempts have been made to distinguish between the two possibilities—failure to fragment versus fragmentation followed by reentry of* $^{57}Fe^{n+}$ *into a neighboring molecule through excited exchange within* 10^{-7} *sec. This has been done by investigating the spectra of labeled complexes diluted in inert matrices and inactive cobalt complexes doped with* $^{57}Co^{++}$.

The absence of fragmentation in highly conjugated ring systems opens up interesting possibilities for the use of Mössbauer emission spectroscopy for systematic study of Vitamin B_{12}, *its analogs, and other biologically important molecules possessing conjugated ring systems, for*

*This work was supported by the U.S. Atomic Energy Commission and NASA Grant No. 237-62.

instance, porphyrins. It can also find applications in the study of bulky iron-bearing molecules where iron can be replaced by cobalt; in such cases, the emission Mössbauer spectroscopy would be decidedly more sensitive than the conventional absorption Mössbauer spectroscopy.

INTRODUCTION

A cobalt-57 nucleus undergoes radioactive decay by capturing an orbital electron—most frequently from the *K* shell. An electron from the less tightly bound *L* shell hops in to fill the vacancy, while the excess energy is utilized in ejecting another electron from the *L* shell of the atom. These two vacancies in the *L* shell are in turn filled by electrons from the next outer shell or subshell. Thus, the vacancy cascades traverse to the valence shell, resulting in a multiply-charged ionic species. Pollak [1] has worked out this Auger cascade scheme for 57-Co^{3+}, and he estimates the maximum charge build up for the daughter iron-57 as +8 (Fe^{8+}). This Auger vacancy cascade is believed [2] to take place in less than 10^{-15} sec. When the cobalt atom is incorporated in a molecule the multiply-charged daughter iron ion is rapidly neutralized by electrons from neighboring atoms in the molecule. The positive charges would be driven to the extremities of the molecule due to electrostatic repulsion. This redistribution of charges in the molecule probably occurs within 10^{-15} sec, and hence without any disturbance in the relative positions of the individual constituent nuclei. In the gaseous phase, where the molecules are isolated, there is considerable experimental evidence to suggest that the multicenter Coulombic repulsion virtually explodes a molecule, forming several fragments [2, 3]. Most of this work has been done with molecules incorporating bromine and iodine atoms, where a large number of electrons are ejected, and there is evidence of violent disruption of molecules forming highly ionized fragments possessing large kinetic energy [2, 3]. For molecules incorporating lighter atoms like cobalt, where fewer Auger electrons are ejected, one would expect the event to be somewhat milder. A similar picture of disruption of a molecule in liquids, with the additional formation of radicals in the immediate vicinity by the Auger electrons has been used by Libby [4], Willard [5], their co-workers, and others. However, little is known about the phenomenon in solids.

For molecules incorporating Co^{57}, the 14.4-keV Mössbauer emission can give us information regarding the chemical forms in which the daughter iron-57 is "stabilized" within 10^{-7} sec following the electron capture. Wertheim *et al.* [6] studied the chemical consequences of electron capture in Co(III) acetylacetonate, and we briefly summarize their interpretation. The highly ionized iron ion undergoes electronic relaxation until it acquires the charge of +3, at which point it is stabilized as Fe(III)$(acac)_3$ in the Co$(acac)_3$ lattice and the sequence is terminated. A charge state of 2+ results if the electronic relaxation permits the transfer of an additional electron to the valence orbitals of the iron atom before the ligands rearrange to accommodate a 3+ charge state. This charge state may then be stabilized, within a time short compared to 10^{-7} sec, as an Fe(II) impurity atom in the Co(III)$(acac)_3$ matrix, or it may be reoxidized by the near-neighbor environment and then stabilized as an Fe(III) atom in a regular lattice site, which differs only slightly from its Co(III) neighbors.

No consideration was given to the possibility of fragmentation of the molecules as a consequence of the Auger event.

In the work reported here we carried out a systematic study of the chemical consequences of Auger ionization in several cobalt chelates labeled with Co^{57}. Fragmentation of the molecule was identified by the presence of degraded high-spin Fe^{2+} and Fe^{3+} species. Parallel experiments with 57-Co^{2+}-doped chelates were conducted to help identification of Fe^{2+} and Fe^{3+} species, as their characteristics, isomer shifts, and quadrupolar splittings are determined by the environment. While the presence of ionic high-spin iron species can be regarded as an unequivocal sign of fragmentation, the reverse may not be true, that is, the presence of coordinated Fe(III) or Fe(II) may reflect failure to fragment or may be a consequence of "reentry" of the degraded Fe^{3+} (resulting from fragmentation) into a neighboring inactive cobalt chelate molecule to form a coordinated molecule. The mechanism of this process will be discussed later. The possibility of exchange can also be examined with the help of doping experiments.

DATA AND DISCUSSION

We would like to mention a few general features of these spectra before discussing individual cases.

In our earlier experiments doping of the chelate was done with carrier-free 57-Co^{2+}. Later it was felt that this procedure does not simulate the environment of the degraded Fe^{2+} species resulting from Auger fragmentation of the parent molecule for the following reasons: When molecular fragmentation occurs due to Coulombic repulsion the iron possesses some kinetic energy and may come to rest in a new intermolecular site, which may be relatively free of "dirty fragments" resulting from the original Auger event. On the other hand, since 57-Co^{2+} is not attached to any coordinating ligand, it is not likely to possess any kinetic energy to move it away from its original site. Hence, it might have a "dirtier" environment, possibly having some fragments in the neighborhood, formed during charge neutralization following the Auger event. To overcome this difficulty, we have used in most cases carrier-free Co(III) citrate as a dopant. From an independent experiment we know that the citrate complex fragments in almost 100% of the Auger events forming ionic Fe^{2+} and Fe^{3+}. This fragmentation is likely to give some kinetic energy to the degraded iron species, which will possibly find itself in an interstitial site in the host lattice. This line of reasoning receives considerable support from our data on doped Co(III) acetylacetonate and Co(III)$(dipy)_3(ClO_4)_3$, to be discussed later.

The degraded high-spin Fe^{2+} species is characterized by a large isomer shift and quadrupole splitting. In some cases the asymmetry of the peaks is very marked. The main reason for this asymmetry appears to be the anisotropy in amplitude of thermal vibration of Fe^{2+} ion as an interstitial in the bulk or adsorbed on the surface. The asymmetry decreases considerably at liquid-helium temperature. However, it is not unlikely that in some cases the spectra of Fe^{3+} ions have a quadrupole component (due to neighboring polar radicals or fragments) superimposed on the Fe^{2+} peak near zero velocity, thereby exaggerating the asymmetry. We have not included this in our curve analysis, as we have no reason to believe that this is the case, nor is it likely to affect the main trend of interpretation of the data. So we have analyzed degraded Fe^{3+} as a single peak centered around ~ 1.2 mm/sec.

Sodium ferrocyanide (1 mg/cm^2 Fe^{57}) has been used as an absorber for all emission experiments, and the shifts are all referred thereto.

The spectra of individual chelates are as follows.

Sodium Co(III) EDTA

Figures 1 and 2 are the spectra of labeled sodium Co^{57}(III) EDTA and the doped chelate, respectively. One can see the quadrupole splittings of Fe^{2+} and the single peak for Fe^{3+}. Peaks corresponding to sodium Fe(III) EDTA [7, 8] would be expected to occur at about -1

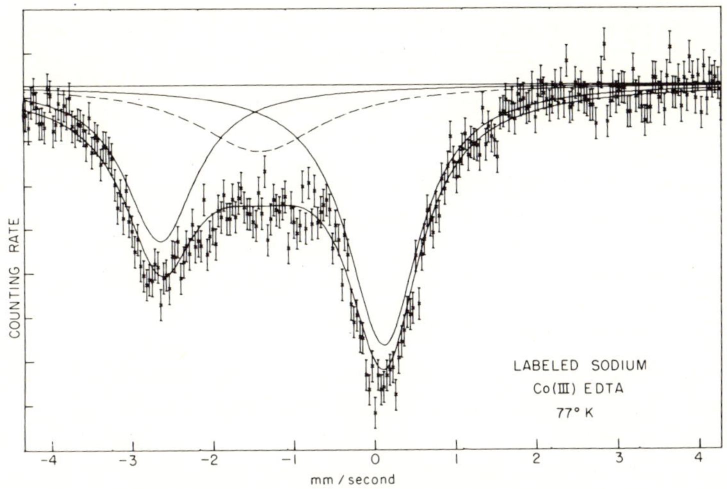

Fig. 1. Emission Mössbauer spectrum of labeled sodium Co^{57}(III) EDTA at 77°K. Solid line, Fe^{2+}; dashed line Fe^{3+}.

and 0 mm/sec, the former being more prominent. It is apparent from Fig. 1 and 2 that if there is any sodium Fe(III) EDTA, it must be present in relatively small amounts. Thus, we can conclude that the EDTA chelate molecule ruptures as a consequence of Auger

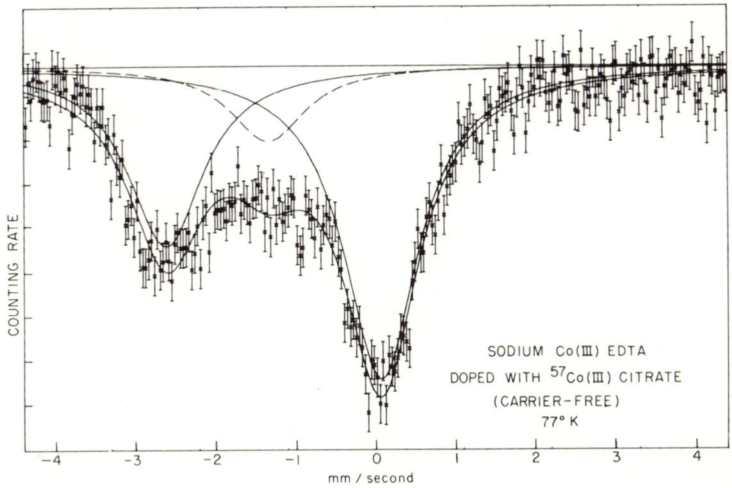

Fig. 2. Emission Mössbauer spectrum of sodium Co(III) EDTA doped with "carrier-free" Co^{57}(III) citrate at 77°K. Solid line Fe^{2+}; dashed line, Fe^{3+}.

ionization in a large majority of events. It may be remarked that this chelate is regarded as being very stable chemically.

Bis-salicylaldehyde-Triethylenetetramine Cobalt(III) Chloride, [Co(III)TS$_2$]Cl

For this complex the situation is similar to the EDTA complex. There is evidence for fragmentation in almost 100% of the events (Figs. 3 and 4). The Fe(III)TS$_2$ · Cl is expected to have peaks at −1.47 and 1.17 mm/sec, and there is no visible indication of their presence.

Bis-indenyl Co(III) Nitrate

Figures 5 and 6 show the spectra of the labeled and doped complex. The spectrum for the labeled bis-indenyl complex bears no resemblance to that of the doped material, unlike the EDTA and TS$_2$ complexes discussed above. Neither does it correspond to the spectrum of the Bis-indenyl Fe(II), where the expected line positions [9] would be −1.82 and 0.72 mm/sec, nor to that of the unstable [bis-indenyl Fe(III)]$^+$ which, on the basis of analogy with [bis-cyclopentadienyl Fe(III)]$^+$ should have a single peak at approximately −0.35 mm/sec [10].

The line is also unusually wide, i.e., 1.73 mm/sec—as a matter of fact, the widest observed for all emission experiments. We would

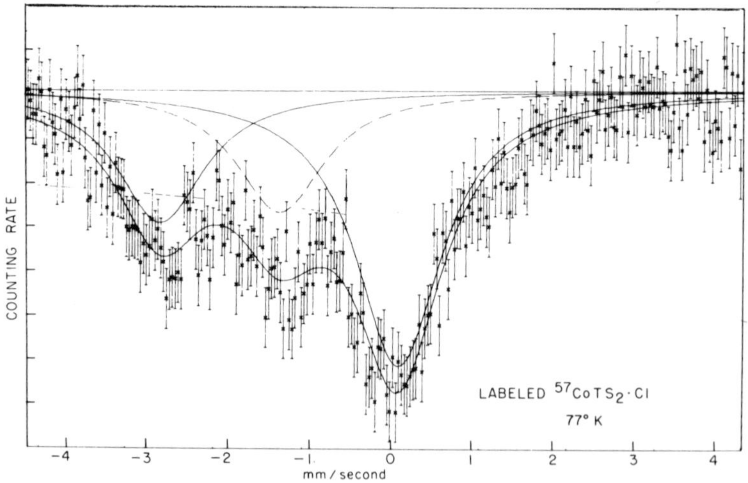

Fig. 3. Emission Mössbauer spectrum of labeled bis-salicylaldehyde triethylenetetramine Co^{57}(III) chloride at 77°K. Solid line, Fe^{2+}; dashed line Fe^{3+}.

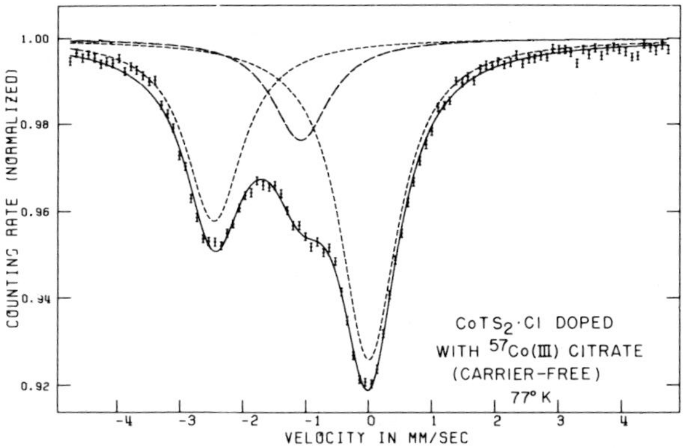

Fig. 4. Emission Mössbauer spectrum of bis-salicylaldehyde triethylenetetramine Co(III) chloride doped with "carrier-free" Co^{57}(III) citrate at 77°K. The outer pair of lines are the quadrupolar split doublet of Fe^{2+}. The central line corresponds to Fe^{3+}.

like to attribute the spectrum to the fragment or radical C_9H_7Fe formed as a result of rupture of one of the rings from the sandwich compound. Hence, in this case also fragmentation occurs as a consequence of Auger ionization.

Cobalt(III) acetylacetonate

Figures 7 and 8 show the spectra of the labeled and doped acetylacetonate complex. With such a close resemblance between the two spectra, one is at first sight led to believe that there is complete

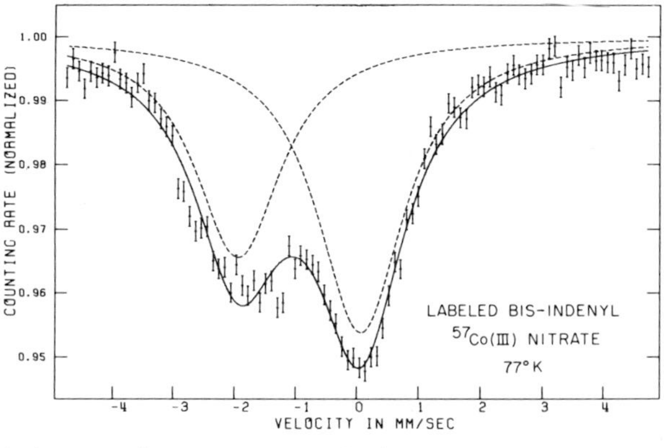

Fig. 5. Emission Mössbauer spectrum of labeled bis-indenyl Co^{57}(III) nitrate at 77°K. The doublet arises from $C_9H_7Fe^+$.

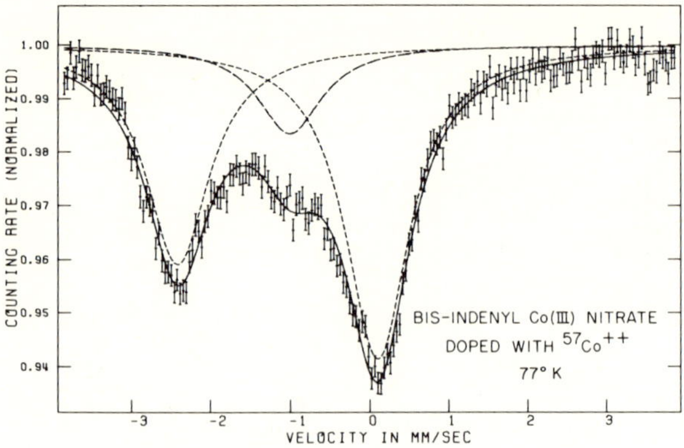

Fig. 6. Emission Mössbauer spectrum of bis-indenyl Co(III) nitrate doped with carrier-free $CoCl_2{}^{57}$ at 77°K. The outer pair of lines are the quadrupolar split doublet of Fe^{2+}. The central line corresponds to Fe^{3+}.

fragmentation resulting in formation of high-spin Fe^{3+} and Fe^{2+} ions. Fe(III) acetylacetonate is expected [11] to show up at ~ -0.5 mm/sec, but with some broadening and distortion due to partial paramagnetic hyperfine splitting as a result of dilution in a diamagnetic matrix, namely, Co(III)(*acac*)$_3$. This region also overlaps with the prominent peak of Fe^{2+}, and hence no attempt was made to analyze the peak (or peaks) for Fe(*acac*)$_3$. We decided rather to detect the presence of

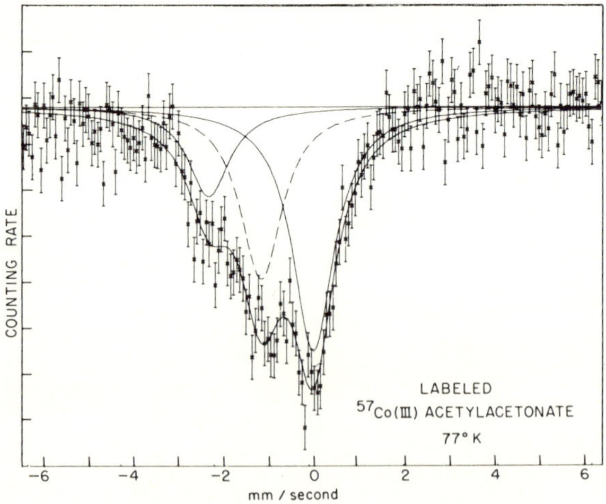

Fig. 7. Emission Mössbauer spectrum of labeled Co^{57}(III) acetylacetonate at 77°K. Solid line, Fe^{2+}; dashed line, Fe^{3+}.

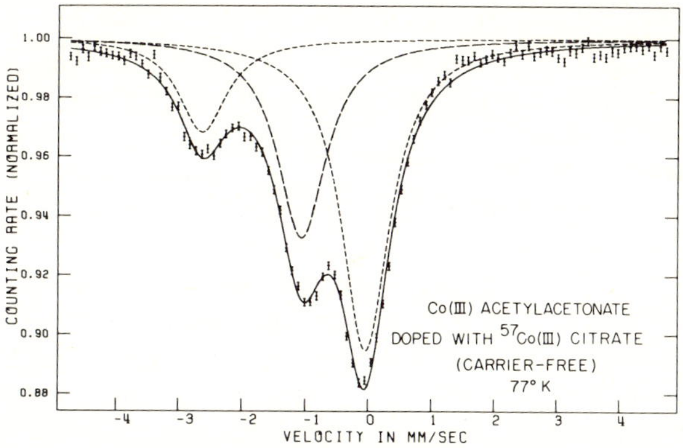

Fig. 8. Emission Mössbauer spectrum of Co(III) acetylacetonate doped with "carrier-free" Co^{57}(III) citrate at 77°K. The outer pair of lines are the quadrupolar split doublet of Fe^{2+}. The central line corresponds to Fe^{3+}.

Fe(*acac*)$_3$ by looking at the positions of the four outer lines at liquid-helium temperature [12]. In Fig. 9 the positions of these lines indeed do correspond with those of Fe(*acac*)$_3$ determined earlier [12] in a diamagnetic dilute solution.

Now the question arises whether the Fe(*acac*)$_3$ originates as a result of failure of molecular fragmentation or reentry following fragmentation, by the following mechanism. A higher charged iron ion, for instance, $\lesssim 4+$, resulting from molecular fragmentation comes to rest at a new site between some Co(*acac*)$_3$ molecules. It

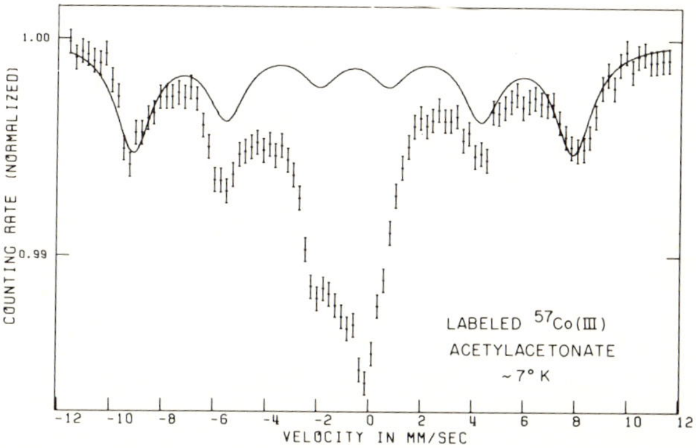

Fig. 9. Emission Mössbauer spectrum of labelled Co^{57}(III) acetylacetonate at about 7°K. The four outer lines of the hyperfine sextet arise from Fe(*acac*)$_3$.

captures an electron to form an electronically exicted Fe^{3+} ion, which in turn can undergo exchange with Co(*acac*)$_3$, forming Fe(*acac*)$_3$. This exchange in solids, excited by electron capture, has received considerable attention [13–15] in a different context. To check whether some fraction of the Fe(*acac*)$_3$ detected is formed through this reentry process, we ran a sample of Co(*acac*)$_3$ doped with carrier-free Co^{57}(III) citrate at liquid-helium temperature (Fig. 10). The result shows clearly the outer four lines of the hyperfine sextet at the expected positions. It may be mentioned that when 57-Co^{2+} is used as the dopant instead of the citrate, these lines are hardly visible. This leads us to believe that the citrate can simulate the conditions of the labeled complex better than Co^{2+}, for reasons discussed earlier.

We anticipated that formation of Fe(*acac*)$_3$ through "reentry" could be prevented by dispersing Co^{57}(*acac*)$_3$ in an inert matrix like cellulose acetate. The spectrum obtained at ~7°K is shown in Fig. 11. The line intensities are much less than observed for the undiluted complex (Fig. 9). One can suggest that this is the fraction of Fe(*acac*)$_3$ resulting from failure to fragment. However, this may not be true. The lattice of the compound is broken down by dispersal in a foreign matrix, and this might affect the rate of charge and energy transfer from the molecule undergoing an Auger event. So the chances of survival of the molecule in a foreign matrix may not be the same as in the parent lattice. Because of these considerations, it is not possible to determine the fraction of Fe(*acac*)$_3$ arising from absence of

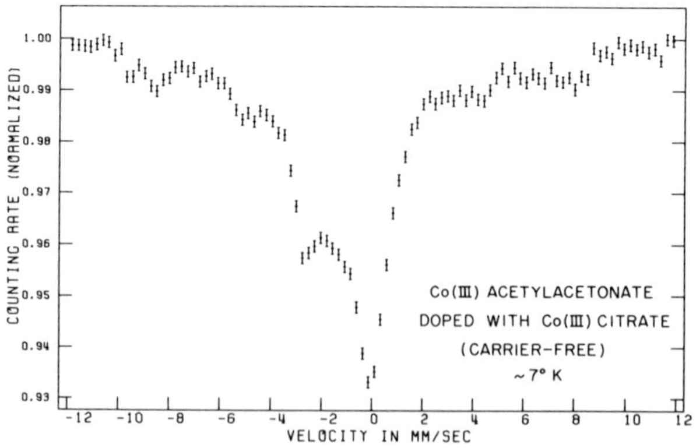

Fig. 10. Emission Mössbauer spectrum of Co(III) acetylacetonate doped with "carrier-free" Co^{57}(III) citrate at about 7°K. The four outer lines of the hyperfine sextet arise from Fe(*acac*)$_3$.

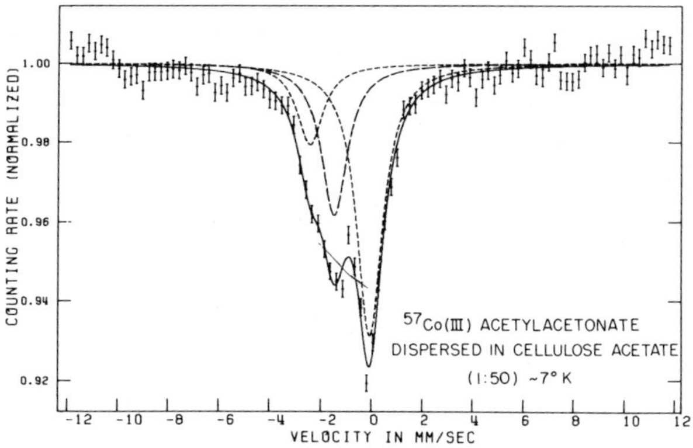

Fig. 11. Emission Mössbauer spectrum of Co^{57}(III) acetylacetonate dispersed in cellulose acetate (1 : 50 by weight) at about 7°K. The four outer lines arising from Fe(*acac*)$_3$ are hardly perceptible.

fragmentation *versus* the fraction of Fe(*acac*)$_3$ resulting through the sequence of fragmentation of the parent molecule and reentry in a neighboring chelate molecule through excited exchange.

Tris-dipyridyl Co(III) perchlorate

The spectra for the labeled and the doped dipyridyl complex are shown in Fig. 12 and 13. In Fig. 12 the two prominent peaks belong

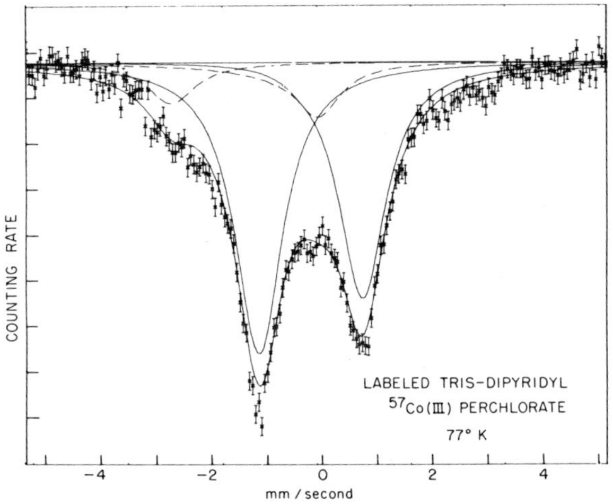

Fig. 12. Emission Mössbauer spectrum of labeled tris-dipyridiyl Co^{57}(III) perchlorate at 77°K. Dashed line, Fe^{2+}, Solid line, Fe(*dipy*)$_3$(ClO_4)$_3$.

to Fe(III)(*dipy*)$_3$ · (ClO_4)$_3$, while the remaining two weak lines constitute the quadrupolar split lines of Fe^{2+}. So the major contribution is from Fe(III)(*dipy*)$_3$(ClO_4)$_3$. From Fig. 13 we learn that a certain fraction of this species arises through "reentry," presumably not a major fraction.

Incidentally, it may be mentioned that if the dipyridyl is doped with carrier-free Co^{57}(III) EDTA, the results are similar. If the dopant is 57-Co^{2+}, there is no evidence for the presence of Fe(III) dipyridyl complex. This point has already been discussed earlier.

Now let us see what happens when we disperse Co^{57}(*dipy*)$_3$(ClO_4)$_3$ in an inert matrix like polyethylene oxide (Fig. 14). In a great majority of events the molecules fragment. It seems that in dilute dispersals not only is the reentry process inhibited, but the probability of survival of the molecule after the Auger event diminishes.

Co(II) Phthalocyanine and Vitamin B_{12}

In Fig. 15 we see no evidence of the presence of the degraded Fe^{2+} species, and the spectrum corresponds to that of Fe(II) phthalocyanine. There is no evidence for any significant excited exchange in the doped compounds, as is to be expected from earlier studies on electronically excited exchange in metal chelates [13–15]. The same

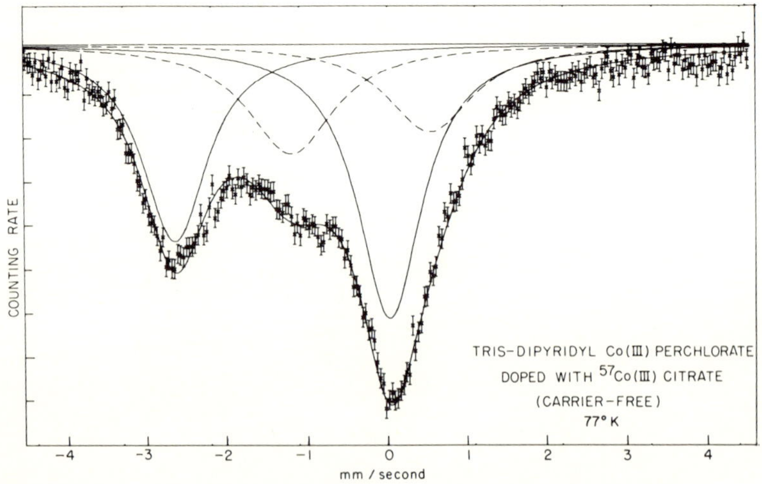

Fig. 13. Emission Mössbauer spectrum of tris-dipyridyl Co(III) perchlorate doped with "carrier-free" Co^{57}(III) citrate at 77°K. Solid line, Fe^{2+}; dashed line, Fe(*dipy*)$_3$(ClO_4)$_3$.

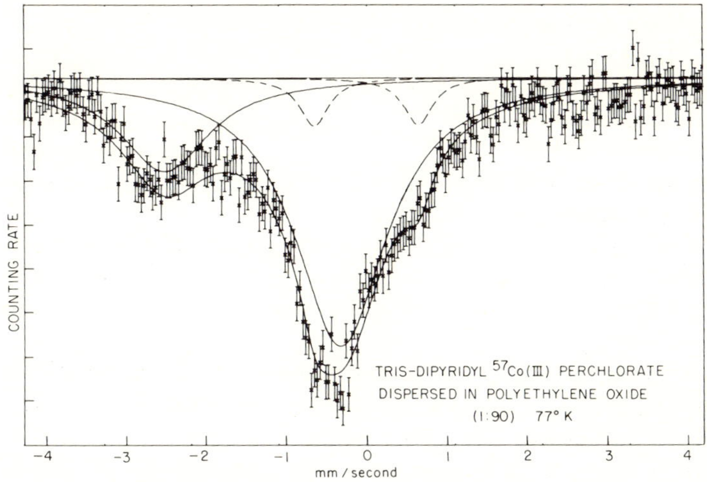

Fig. 14. Emission Mössbauer spectrum of tris-dipyridyl Co^{57}(III) perchlorate dispersed in polyethylene oxide (1 : 90 by weight) at 77°K. Solid line, Fe^{2+} and Fe^{3+}; dashed line, Fe(*dipy*)$_3$(ClO_4)$_3$.

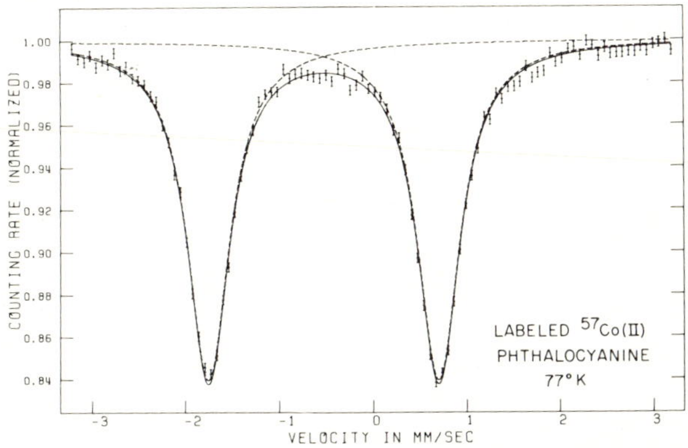

Fig. 15. Emission Mössbauer spectrum of labeled Co^{57}(II) phthalocyanine at 77°K. The doublet corresponds to Fe(II) phthalocyanine.

picture obtains for vitamin B_{12}, as is evident from Mullen's work [15a], as well as ours [16].

CONCLUDING REMARKS

The following picture emerges from these studies:

1. Cobalt chelates with hexadentate ligands like EDTA and TS_2, which are chemically very stable, fragment in a large majority of

events as a consequence of the Auger event. The same is true for the indenyl complex.

For the system Co(III)(*acac*)$_3$ the spectra reveal a small fraction of coordinated iron. This [i.e., Fe(*acac*)$_3$] arises by virtue of two processes: the parent molecule escaping fragmentation, and the "reentry" of the degraded iron ion undergoing excited exchange with a host cobalt chelate molecule.

For the system Co(III)(*dipy*)$_3$(ClO_4)$_3$ a majority of events do not result in fragmentation, while a small contribution from reentry is likely. Dispersal of the dipyridyl complex in an inert matrix enhances the probability of fragmentation.

There is no evidence of fragmentation in Co^{57}(II) phthalocyanine or in vitamin B_{12}. From Fig. 16 it is apparent that in phthalocyanine the metal atom is coupled to a large and highly conjugated system (unlike EDTA, acetylacetone, and TS_2 complexes—Figs. 17 and 18), and the same is true of vitamin B_{12}. One can visualize that the multiply-charged Fe^{57} ion extracts electrons from the ring, and the resulting positive charges on the ring are dispersed to its periphery. Due to the rapid dispersal of charge through the conjugated system and due to the large size of the ring [18] the multicenter Coulombic repulsion

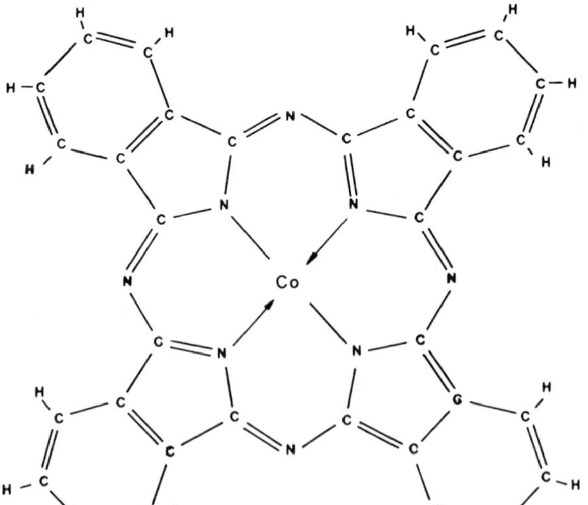

Fig. 16. Structural formula of Co(II) phthalocyanine.

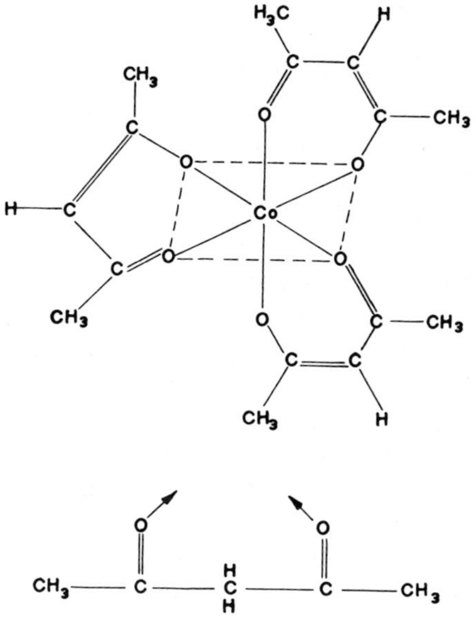

Fig. 17. Structural formula of Co(III) acetylacetonate.

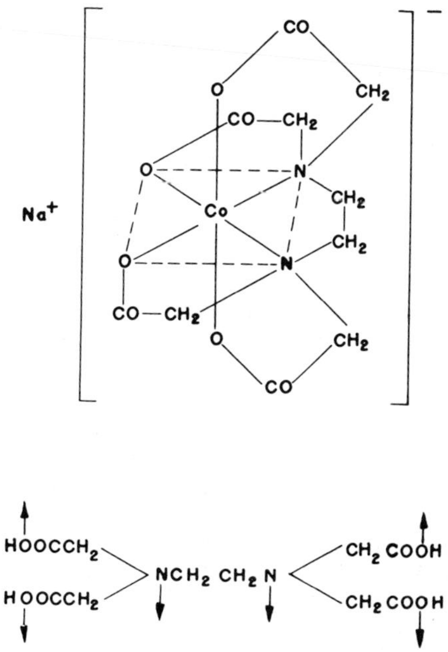

Fig. 18. Structural formula of sodium Co(III) EDTA.

is inadequate to fragment the molecule. These charges on the ring are apparently rapidly dispersed through neighboring molecules before the bonds get a chance to miss the electrons, namely, within 10^{-13} sec. The rapid charge dispersal through neighboring molecules is perhaps also facilitated as the relative positions of atoms remain fixed during this interval, and Franck–Condon restrictions would not arise. As a consequence of charge neutralization, one would expect several tens of electron volts of energy to be deposited in the parent molecule and its immediate vicinity. It is amazing that this large amount of energy is so efficiently dispersed as to leave this molecule unfragmented. This picture appears to be true to a large extent for Co(III)$(dipy)_3$ $(ClO_4)_3$. When the dipyridyl molecules are dispersed in an inert matrix the original lattice is broken down. The dissipation of charge and energy is no longer rapid and efficient, and thus the probability of fragmentation is enhanced.

It may be pertinent to mention that Herber and Stöckler's [17] observations on tetraphenyl tin are also suggestive of absence of fragmentation.

2. There is experimental evidence to suggest that a certain fraction of the coordinated iron chelate can, in cases of the acetylacetone and dipyridyl complexes, arise from reentry of the multiply-charged degraded iron ion.

3. The isomer shifts and quadrupolar splittings of the ferric chelates in these emission experiments sometimes differ from the values determined through absorption experiments. For instance, both these values for Fe(III)$(dipy)_3(ClO_4)_3$ in the labeled cobalt chelate and doped sample differ from those obtained for the dispersed sample. Those obtained through absorption measurements [7] in turn, differ from either of these emission experiments. It appears that slight distortion of Fe(III)$(dipy)_3$ occurs in matrices of Co(III)$(dipy)_3 \cdot ClO_4$, and polyethylene oxide. Similar differences between absorption and emission spectroscopy have been observed for phthalocyanines [18, 19].

ACKNOWLEDGMENTS

Two of the authors (A.N. and D.L.) are particularly grateful to Dr. W. F. Libby for several valuable discussions. The authors would also like to express their thanks to Mr. Wallace Erwin and Miss Lucy Marton for their help during the course of the work.

REFERENCES

1. H. Pollak, *Phys. Stat. Sol.* **2**:720 (1962).
2. S. Wexler, in: *Actions Chemiques et Biologiques des Radiations*, M. Haissinsky, ed. (Masson, Paris, 1965), Vol. 8, pp. 103–241.
3. A. H. Snell, *α, β, and γ-Spectroscopy*, K. Siegbahn, ed. (Interscience, New York, 1964), p. 1545.
4. A. R. Kazanjian and W. F. Libby, *J. Chem. Phys.* **42**:2778 (1965).
5. P. R. Geissler and J. E. Willard, *J. Phys. Chem.* **67**:1675 (1963).
6. G. K. Wertheim, W. R. Kingston, and R. H. Herber, *J. Chem. Phys.* **37**:687 (1962).
7. L. M. Epstein, *J. Chem. Phys.* **36**:2731 (1962).
8. J. J. Spijkerman, L. H. Hall, and J. L. Lambert, *J. Am. Chem. Soc.* **90**:2039 (1968).
9. R. H. Herber, R. B. King, and G. K. Wertheim, *Inorg. Chem.* **3**:101 (1964).
10. G. K. Wertheim and R. H. Herber, *J. Chem. Phys.* **38**:2106 (1963).
11. G. M. Bancroft, A. G. Maddock, W. K. Ong, R. H. Prince, and A. J. Stone, *J. Chem. Soc. A*, p. 1966 (1967).
12. J. W. G. Wignall, *J. Chem. Phys.* **44**:2462 (1966).
13. J. Shankar, Amar Nath, and M. H. Rao, *Radiochim. Acta* **3**:26 (1964).
14. Amar Nath, S. Khorana, P. K. Mathur, and S. Sarup, *Indian J. Chem.* **4**:51 (1960).
15. Amar Nath and S. P. Vaish, *J. Chem. Phys.* **46**:4660 (1967).

15a. R. T. Mullen, this volume.

16. Amar Nath, Michael Harpold, and Melvin P. Klein, *Chem. Phys. Letters* **2**:471 (1968).
17. R. H. Herber and H. A. Stöckler, in: *Proceedings of the Symposium on Chemical Effects of Nuclear Transformations*, I.A.E.A., Vienna (1965), Vol. II, p. 403.
18. J. L. Thompson, private communication.
19. Amar Nath, Melvin P. Klein, and Walter Kündig, unpublished work.

Methodology

Mössbauer Effect in Radioactive Materials*

A. J. F. Boyle† and G. J. Perlow

Argonne National Laboratory
Argonne, Illinois

The general considerations that apply to using unstable isotopes as the absorbing nuclei for the Mössbauer effect are discussed. Preliminary results of various workers with the isotopes 2.6-year Pm^{147}, 7.2-year Ba^{133}, 87-year Sm^{151}, *and* 47-hr Sm^{153} *are discussed as examples of the techniques involved.*

INTRODUCTION

During the ten years since its discovery the Mössbauer effect has been observed in some 50 gamma-ray transitions, and has yielded valuable information on the properties of the nuclear states involved, as well as on many solid-state and chemical properties of the host materials of the emitting and absorbing nuclei.

It is usually considered a *sine qua non* of the Mössbauer effect to have a stable or almost stable isotope as the resonant absorber (or scatterer) so that at least milligram quantities may be used. On the other hand, this is no compelling reason why the effect should not be observable in short-lived isotopes, since the resonant cross sections are large, and therefore relatively few nuclei per unit area are needed. Extension of the technique to include such isotopes would greatly expand the field of nuclear investigation and make some new elements available for solid-state and chemical studies.

In this article we shall first discuss the general considerations appropriate to the use of radioactive absorbers, and follow with a brief description of experiments and results obtained with four such

*Based partly on work performed under the auspices of the US Atomic Energy Commission.

†On leave from the University of Western Australia.

absorbers: 2.6-year Pm^{147}, 7.2-year Ba^{133}, 87-year Sm^{151}, and 47-hr Sm^{153}.

GENERAL CONSIDERATIONS

The prime consideration is simply to arrange for the number of absorbing nuclei per unit area to be sufficient for the observation of an effect. A limit is set by the available specific activity of the absorber material and the optimum total thickness of the absorber as determined by atomic absorption. The calculation of this optimum must include the contribution to the background in the counting channel that arises from the absorber activity. This optimum will depend on the particular case, but, in general, one atomic absorption length is appropriate. A value somewhat less than this would be desirable if the absorber background tended to dominate the channel, but in this event a means of increasing the source strength would first be sought. In many cases the specific activity is limited if the absorber must be produced by neutron irradiation. With present available neutron flux levels this precludes many otherwise possible transitions—particularly among the shorter-lived isotopes.

The relative contribution of the absorber to the counting channel can be reduced by placing it close to the source. The increase in Rayleigh scattering of the source radiation introduced by this close placement will not usually be significant. The area of the source can then be nearly equal to that of the absorber, but the thickness of the source is also limited effectively to one or two atomic absorption lengths. Thus, for a reasonable signal-to-background ratio the *specific* activity of the source should be comparable to that of the absorber (insofar as the latter contributes to the counting channel). In the case of sources produced by neutron irradiation, or even for carrier-free material of longer half-life than the absorber, this condition may prove difficult to meet.

The areas of both source and absorber are then fixed by one of the following conditions: (1) The total absorber activity may be limited either by reason of difficulties in handling the active material, or by its cost. The accompanying limit on total source activity may result in an excessively low counting rate, and indicate the need for special measures to increase the specific activity of the source. (2) A limited total source activity, even if carrier free, may control the permissible total absorber activity, and hence the areas. (3) A maximum permissible counting rate may determine the maximum areas with the detector at its minimum distance.

2.6-year Pm^{147}

Croft, *et al.* [1] were the first to use a short-lived, and hence highly radioactive, isotope as a Mössbauer absorber when they observed an effect with the 91-keV transition in Pm^{147}.

The partial decay scheme of the source and the absorber is shown in Fig. 1. With this absorber it is possible to shield the detectors from the bremsstrahlung and the fluorescence X-ray produced in its decay without severe attenuation of the gamma ray of interest. Also, Pm^{147} is a fission product and is available, virtually carrier free, in gram quantities. For these reasons, therefore, there was no need to depart from the conventional experimental arrangement in which the absorber area is around 1 cm^2 and 10 mg ($\sim$10 Ci) or more of absorber material is used.

Figure 2 shows a typical absorption spectrum obtained by Croft *et al.* Several chemical (trivalent) forms of the absorber were examined, and no isomer shifts were found within the experimental accuracy ($\pm$0.7 mm/sec). The experimental linewidth was 4.0 $\pm$ 0.7 mm/sec, which is three times the natural width. The size of the effect ranged up to near 0.5%.

7.2-year Ba^{133}

Figure 3 shows the decay scheme of the absorber and the source, which in this case is an isomeric state of the absorber nucleus.

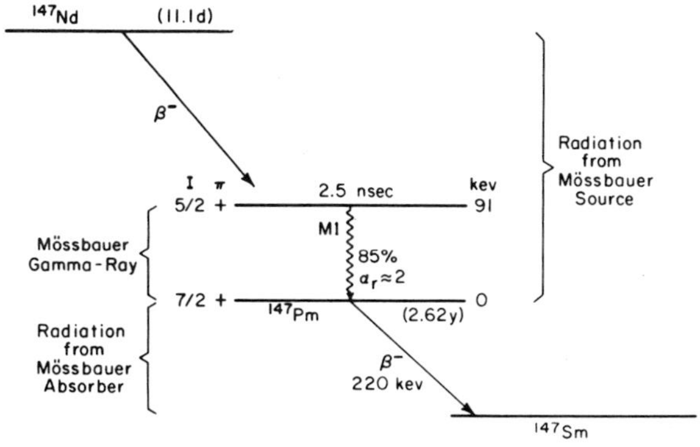

Fig. 1. Partial decay scheme of Nd^{147} and Pm^{147}.

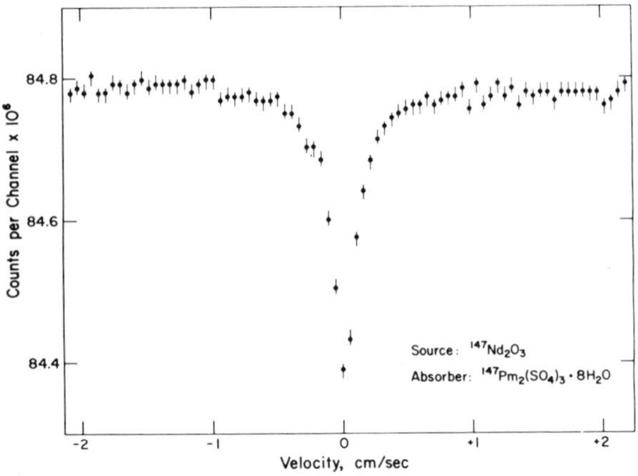

Fig. 2. Typcial velocity spectrum for the 91-keV transition in Pm^{147}.

Ba^{133} can be obtained with a specific activity of 6 Ci/g. One atomic absorption thickness for the 12.3-keV gamma ray is 10 mg/cm^2 so the "thickness" $t = nf\sigma$ is ~ 0.25.

In the counting channel the principal background arising from the absorber comes from the 30-keV X-ray produced in the decay by electron capture. This can be minimized by the use of an argon-methane proportional counter, but still demands a certain specific activity per unit area for the source. This is difficult to achieve by neutron irradiation of Ba^{132}. The Ba^{133} was produced with the $Cs^{133}(d,2n)$ reaction, so that the activity can be extracted more or less carrier-free. This procedure results in a limited total source activity and a correspondingly limited absorber area.

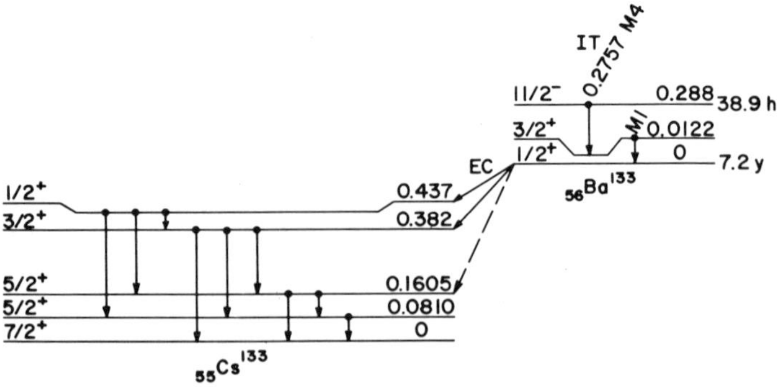

Fig. 3. Partial decay scheme of Ba^{133m} and Ba^{133}.

The source activity was extracted by adding 50 μg of calcium ion to the solution of the CsCl target and precipitating the carbonate; this was then usually converted to the chloride. The source formed a thin disk 1.2 mm in diameter. The absorber, also usually the chloride, had a diameter of 1.5 mm and contained 1 mCi (2 μg) of active material. The source moved 0.25 mm about a mean separation of 0.5 mm between source and absorber. The apparatus was constructed so that the necessary close tolerance in alignment could be obtained and preserved.

Figure 4 shows a velocity spectrum obtained in one source half-life with the source and absorber at 4.2°K. No isomer shifts were found with different chemical forms (oxide and sulphate) of the source. The experimental linewidth in all cases was 8 mm/sec, which is nearly three times the expected natural width. This broadening cannot be attributed to unresolved quadrupole effects, since in one case the source (and absorber) were in the cubic form of BaO.

Some additional work with this isotope serves to illustrate the routine nature of the experimentation when the miniaturized set up is used. Figure 5 shows the velocity spectrum obtained with the source and absorber in a longitudinal field of 76.30 kOe provided by a superconducting magnet. In this case of a (3/2) $\rightarrow$ (1/2) transition the energy separation of the outer peaks is $2\Delta gH$, from which we can deduce that the difference Δg between the algebraic gyromagnetic ratio of the ground state and that of the excited state is 1.98 $\pm$ 0.03 nm. Since the ground-state moment is not known, a further experiment either with a transverse field or with a shielded source (or absorber)

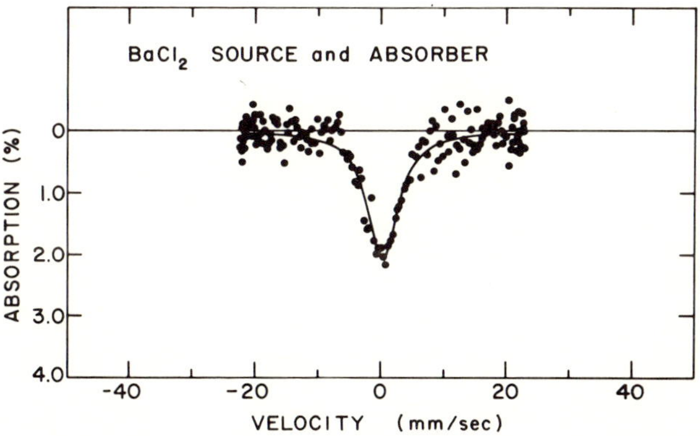

Fig. 4. Typical velocity spectrum for the 12.3-keV transition in Ba^{133}.

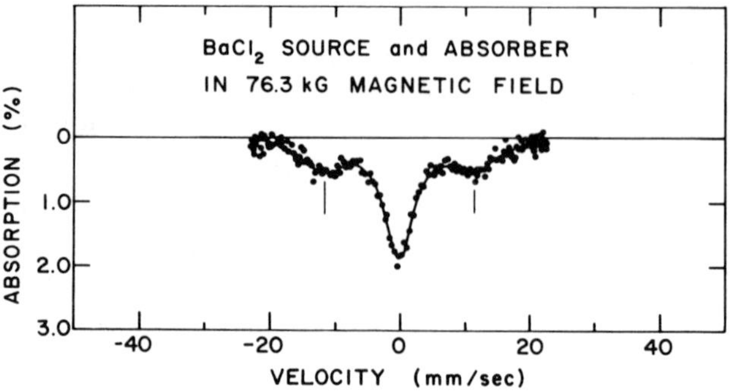

Fig. 5. Velocity spectrum for the 12.3-keV transition in Ba^{133} with source and absorber in a magnetic field of 76.30 kOe.

will be necessary to determine the individual ratios. Nuclear systematics for these spin values in this nuclear region do, however, indicate a value of 2.0 nm for Δg.

87-year Sm^{151} *

Preliminary experiments with this isotope again illustrate the usefulness of the miniaturized system. Figure 6 shows a partial decay scheme for the source and absorber. There are several transitions which might be examined, but the initial work was done with the 65.8-keV line, which has an intensity of $\sim 1\%$.

Sm^{151} is available with a specific activity of 23 Ci/g, and for one atomic absorption length $t \approx 9$. The source is prepared by neutron irradiation of Nd^{150}, and there is no difficulty in obtaining sufficient specific activity to provide a dominant signal in the detection channel.

Fig. 6. Partial decay scheme of Pm^{151} and Sm^{151}.

*Work performed by one of the authors (A.J.F.B.) and G. M. Kalvius.

The prime consideration in this case was that the absorber material is expensive, and only 10 mCi were on hand for these initial experiments. It would have been desirable to have used an absorber area of 1 mm^2, but this would have restricted the counting rate; the compromise adopted was an area of 3 mm^2 and $t = 1.3$. The same apparatus as had been used for the experiments with Ba^{133} was used again here.

Figure 7 shows the velocity spectrum obtained with an oxide source and absorber. The spin of the ground state and of the excited state and the lifetime of the transition are all unknown. The experimental linewidth is about 20 mm/sec, which would correspond to a lifetime of 0.2 nsec for the transition if there are no broadening effects. In fact, since the anticipated effect was about 1–2%, and the observed effect only 0.3%, it is likely that quadrupole effects are important. An estimate of the energy of the quadrupole coupling for this nucleus in the noncubic environment of the oxide suggests a value around 10 mm/sec. Further work with cubic forms will be necessary before any information can be gained on the spins of the levels involved.

47-hr Sm^{153} *

Figure 8 shows the energy-level diagram of Sm^{153} and the decay of the ground state to Eu^{153}. The 35.8-keV transition is not populated

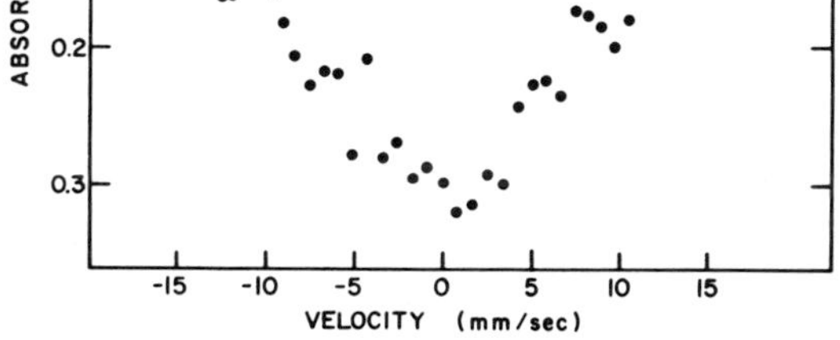

Fig. 7. Velocity spectrum for the 65.8-keV transition in Sm^{151}.

*Work performed by the authors in collaboration with L. M. Bollinger, R. K. Smither, and G. E. Thomas.

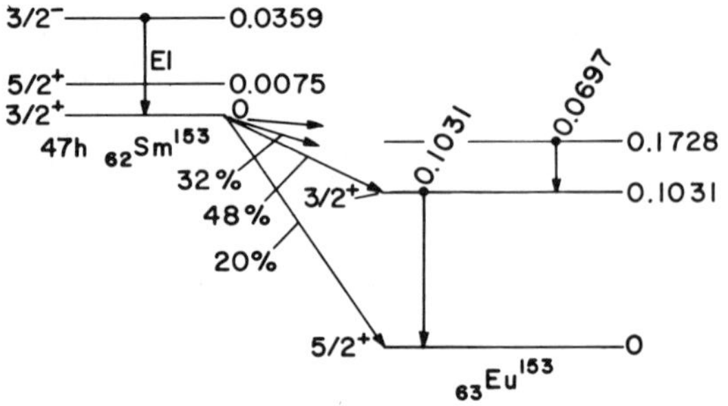

Fig. 8. Partial energy-level diagram and decay scheme of Sm^{153}.

in any radioactive decay and must be produced by the neutron-capture process in Sm^{152}. This fact introduces some new procedures which are relevant not only to unstable absorbers, but also to a number of stable isotopes that are shielded or for which the only source has a very short half-life.

The first consideration is, as usual, the number of absorbing atoms per unit area. Irradiation to saturation of 140 mg/cm^2 of Sm_2O_3 in a flux of 3×10^{13} neutrons/cm^2/sec ($\sigma_n \approx 150$ b) gives about 4×10^{17} atoms of Sm^{153}/cm^2. At 77°K, where $f \approx 0.4$, the Mössbauer "thickness" was about 0.15.

For a stable or weakly active absorber, for which the specific activity per unit area of the source is irrelevant, the only further consideration is the counting rate, and in a few cases [2] it has been possible to produce the source transition with a neutron beam external to the reactor. In the present case, and in most cases of a short-lived absorber, the specific activity per unit area of the source cannot reach the level necessary to dominate the background in the counting channel if it is placed in the external beam. In order to reach such a level, the irradiation time to produce the absorber activity would need to be very short; its ratio to the lifetime of the activity could not be more than 10^{-4} or so (the ratio of external to internal neutron flux).

With the source in the central flux of the reactor the transition of interest is accompanied by the intense general radiation present in the direct beam from the reactor, but can be distinguished with the help of a crystal reflection.

Figure 9 is a schematic diagram of the experimental arrangement. The source was again 140 mg/cm^2 of $^{152}Sm_2O_3$. To minimize random relative motion, the vibrator (with the absorber attached) and the source were clamped rigidly at opposite ends of an aluminum

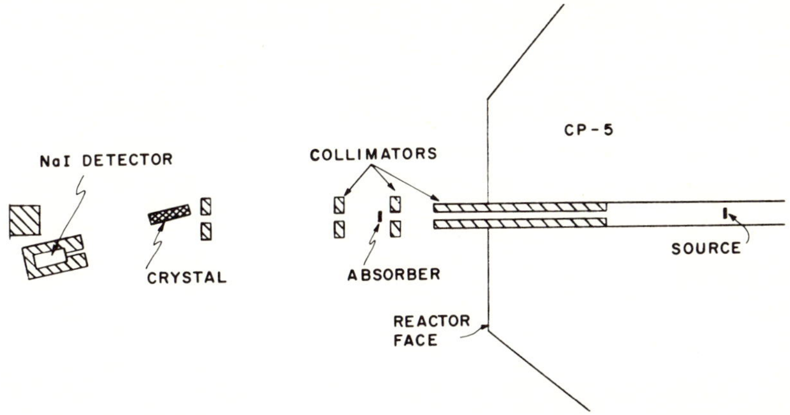

Fig. 9. Schematic diagram of experimental layout.

tube which extended to the center of the reactor. Both the source and absorber were cooled to liquid-nitrogen temperature.

The Cu crystal had a rocking curve of about 20 min and a reflectivity of 10 or 20% for the 35.8-keV gamma ray. The area of the source and hence the counting rate are defined by the collimation needed to benefit fully from the resolution of the crystal. The background in the counting channel included a 17% contribution from the source and 38% from the fresh absorber. The counting rate was around 170/sec.

The velocity spectrum shown in Fig. 10 is a composite of two runs, each of around 50 hr duration. A novel feature of this experiment is that the intensity of the effect diminishes with the decay of

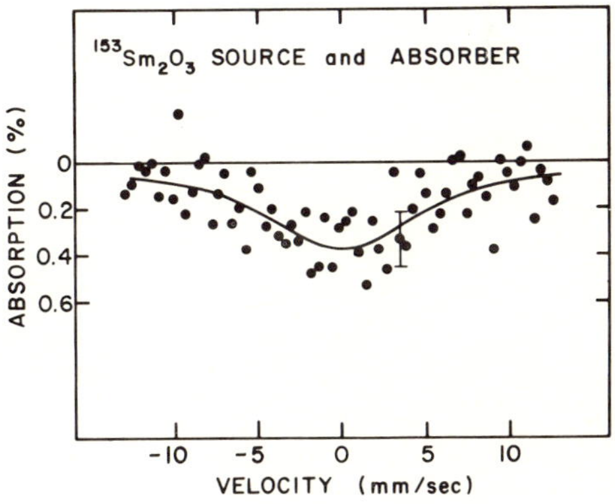

Fig. 10. Velocity spectrum for the 35.8-keV transition in Sm^{153}.

Table I. Mössbauer Absorbers with Half-Lives Less Than 10^2 Years and the Sources Appropriate for Their Use

Absorber						Source	
			Spin				
Nuclide	E_γ, keV	τ_γ, nsec	Ground	Excited	Half-life*	Nuclide	Half-life*
As^{73}	66	5.0	$3/2^-$	$5/2^-$	80 d	Se^{73}	7 h
Rb^{83}	40	—	$5/2^-$	$3/2^-$	83 d	Sr^{83}	32 h
Ba^{133}	12.3	8	$1/2^+$	$3/2^+$	7.2 y	Ba^{133m}	37 h
Cs^{134}	10	47	4^+	5^+	2.05 y	Cs^{134}	2.9 h
Pr^{143}	57	4	$7/2^+$	$5/2^+$	13.6 d	Ce^{143}	33 h
Pm^{145}	61	2.7	$(5/2^+)$	—	17.7 y	Sm^{145}	340 d
Pm^{147}	91	2.6	$7/2^+$	$5/2^+$	2.6 y	Nd^{147}	11 d
Sm^{151}	4.8	2.0	—	—	} 87 y	Pm^{151}	28 h
	66	—	—	—			
	105	0.5	—	—			
Eu^{152}	90	400	3^-	4^+	12 y	Eu^{152m}	96 m
Gd^{153}	42	—	$3/2^+$	$5/2^+$	242 d	Tb^{153}	55 h
Eu^{155}	75	—	$(5/2^+)$	$(7/2^+)$	1.81 y	Sm^{155}	23 min
Eu^{156}	22.6	—	0^+	1^+	15 d	Sm^{156}	9.4 h
Tb^{157}	61	—	$(3/2^+)$	$(5/2^+)$	150 y	Dy^{157}	8.1 h
Yb^{169}	71	—	$(7/2^+)$	$(9/2^+)$	32 d	Lu^{169}	34 h
Tm^{171}	5	—	$1/2^+$	$3/2^+$	1.92 y	Er^{171}	7.5 h
Lu^{174}	45	—	1^-	2^-	3.6 y	Lu^{174m}	140 d
Os^{185}	37	—	$1/2^-$	$3/2^-$	94 d	Ir^{185}	14 h
Au^{195}	61	2.8	$3/2^+$	$(1/2^+)$	183 d	Hg^{195m}	40 h
Ra^{223}	50	0.6	$1/2^+$	$3/2^-$	11.4 d	Th^{227}	18 d
Ac^{227}	27	40	$3/2^+$	—	21.6 y	Pa^{231}	3×10^4 y
						Ra^{227}	41 min
Th^{228}	58	0.4	0^+	2^+	1.9 y	Pa^{228}	22 h
U^{232}	48	0.25	0^+	2^+	72 y	Pa^{232}	1.3 d
						Pu^{236}	2.9 y
Pu^{238}	44	0.2	0^+	2^+	86 y	Cm^{242}	163 d
						Np^{238}	2.1 d
Cm^{242}	42	—	0^+	2^+	163 d	Am^{242}	16 h

*y: years; m: months; d: days; h: hours; min: minutes.

the absorber; this was confirmed by the analysis of the data accumulated during the first 10 hr of the run. Under these circumstances the statistical accuracy of the data reaches a maximum after a running time of around one mean life of the absorber.

With absorbers having very short lives, a few hours or less, it would probably be necessary to have the absorber remain in the central flux so that the absorbing nuclei could be produced continuously along with the source transition.

Since the lifetime of the transition is not known, the experimental linewidth (10 mm/sec) cannot be compared with the natural linewidth; however, an estimate of the effect to be expected suggests either that the line is broadened by a factor of about three, or that a quadrupole pattern would be apparent with better statistics. The most that can be said from the preliminary experiment is that the lifetime of this $E1$ transition is probably between 1 and 3 nsec.

Table II. Mössbauer Absorbers with Half-Lives Longer than 10^2 Years and the Sources Appropriate for Their Use

Absorber						Source	
			Spin				
Nuclide*	E_γ, keV	τ_γ, nsec	Ground	Excited	Half-life, years	Nuclide	Half-life†
[4] I^{129}	27.8	15	$7/2^+$	$5/2^+$	1.7×10^7	Te^{129m}	34 d
La^{137}	10	89	$7/2^+$	$5/2^+$	6×10^4	Ce^{137}	9 h
Pt^{193}	12.6	2.2	—	—	~500 y	Pt^{193m}	4.3 d
	112	—				Au^{193}	16 h
Bi^{208}	63		(5^+)	(4^+)	3.7×10^5	Po^{208}	2.9 y
Ra^{226}	67.8	0.63	0^+	2^+	1.6×10^3	Ac^{226}	29 h
Th^{229}	42	—	$5/2^+$	$7/2^+$	7.3×10^3	Pa^{229}	1.5 d
[5] Th^{230}	53	0.37	0^+	2^+	8×10^4	Pa^{230}	17.7 d
						U^{234}	2.5×10^5 y
Pa^{231}	84	41	$3/2^-$	$5/2^+$	3.2×10^4	Th^{231}	25 h
U^{233}	40	0.37	0^+	2^+	1.6×10^5	Pa^{233}	27 d
[6] U^{234}	44	0.27	0^+	2^+	2.5×10^5	Pu^{238}	86 y
[6] U^{236}	45	0.23	0^+	2^+	2.4×10^7	Pu^{240}	7×10^3 y
[7] Np^{237}	59.5	63	$5/2^+$	$5/2^-$	2.1×10^6	Am^{241}	432 y
						Pu^{237}	46 d
						U^{237}	6.7 d
[8] U^{238}	447	0.23	0^+	2^+	4.5×10^9	Pu^{242}	3.8×10^5 y
Pu^{239}	57	—	$1/2^+$	$5/2^+$	2.4×10^4	Cm^{243}	32 y
						Np^{239}	2.4 d
Pu^{240}	43	0.17	0^+	2^+	6.6×10^3	Cm^{244}	17.6 y
Pu^{242}	44.5	—	0^+	2^+	3.8×10^5	Cm^{242}	5.5×10^3 y
—	—	—	—	—	—	Am^{242}	16 h
[6] Am^{243}	84	2	$5/2^-$	$5/2^+$	8×10^3	Pu^{243}	5 h
Pu^{244}	45	—	0^+	2^+	8×10^7	Cm^{248}	4.7×10^5 y
Cm^{245}	55	—	$7/2^+$	$9/2^+$	9.3×10^3	Cf^{249}	360 y
Cm^{246}	43	—	0^+	2^+	5.5×10^3	Cf^{250}	13 y
Cm^{248}	43	—	0^+	2^+	4.7×10^5	Cf^{252}	2.7 y

*Reference citation (in brackets) preceding an isotope indicates that it has already been studied with a conventional arrangement.
†y: years; d: days; h: hours.

Table III. Unstable Isotopes in Which the Mössbauer Effect Might be Observed and for Which the Source of the Transition is the (n,γ) Reaction

Isotope	E_γ, keV	τ_γ, nsec	Spin: Ground	Spin: Excited	$T_{1/2}$*	σ_n, barns
K^{40}	29	3.9	4^-	3^-	1.7×10^7 y	0.6
Se^{46}	51	—	4^+	—	84 d	10
Mn^{54}	60	—	3^+	—	303 d	170
Mn^{56}	26	11	3^+	2^+	2.6 h	13
Ga^{72}	15	—	3^-	(2)	14 h	5
Nb^{94}	59	—	6^+	(4^+)	2×10^4 y	1
Sb^{122}	61	1800	2^-	3^+	2.8 d	6
La^{140}	30	0.5	3^-	2^-	50 hr	9
Dy^{159}	57	—	$3/2^-$	$5/2^-$	144 d	100
Ho^{166}	54	—	0^-	2^-	27 h	64
Ta^{182}	99	—	(3^-)	—	115 d	21
W^{181}	113	—	$(9/2^+)$	$(11/2^+)$	140 d	20
W^{185}	60	—	$3/2^-$	$(5/2^-)$	75 d	2
Re^{188}	63	—	1^-	2^-	17 h	70
Os^{185}	37	—	$1/2^-$	$3/2^-$	94 d	200
Ir^{194}	43	—	1^-	$(0,1)^-$	17 h	110

*y: years; d: days; h: hours.

CONCLUSION

Tables I–III list those radioactive isotopes for which the observation and use of the Mössbauer effect seems feasible. Several isotopes (indicated by reference citations) have already been studied with a conventional arrangement.

It should be mentioned that although designed for work with short-lived absorbers, the miniaturized set up will often be of benefit when only small amounts of absorber material (radioactive or stable) are available.

The experimental technique used in the observation of the effect in Sm^{153}, although only marginally successful because of the scarcity of absorbing nuclei, will probably be more useful, at least initially, in work with stable absorbers.

REFERENCES

1. W. L. Croft, J. A. Stone, and W. L. Pillinger, *Bull. Am. Phys. Soc.* **12**:565 (1967).
2. D. W. Hafemeister and E. B. Shera, *Phys. Rev. Letters* **14**:593 (1965).
3. J. Fink and P. Kienle, *Physics Letters* **17**:326 (1965).

4. D. W. Hafemeister, G. DePasquali, and H. DeWaard, *Phys. Rev.* **135**:B1089 (1964).
5. N. Hershkowitz, C. G. Jacobs Jr., K. A. Murphy, *Physics Letters* **27B**:563 (1968).
6. B. D. Dunlap, G. M. Kalvius, S. L. Ruby, G. K. Shenoy, D. Cohen, and M. B. Brodsky, private communication.
7. W. L. Pillinger and J. A. Stone, in: *Mössbauer Effect Methodology*, Vol. 4, I. J. Gruverman, ed. (Plenum Press, New York, 1968), p. 217.
8. S. L. Ruby, G. M. Kalvius, D. B. Dunlap, G. K. Shenoy, D. Cohen, and M. B. Brodsky, *Physics Letters* **29B**(8) (1969).

The Mössbauer Effect: A New Method for Measuring Diffusion*

J. G. Mullen and R. C. Knauer†

Physics Department
Purdue University
Lafayette, Indiana

We have been able to directly observe atomic diffusion in solids by observing the broadening of the Mössbauer line. There appears to be some advantage to the technique over the more common tracer-sectioning technique for measuring diffusion. In addition to the fact that the technique does not require the destruction of the sample, thereby permitting the measurement of the temperature dependence of the diffusivity with a single specimen, it may afford more accurate results for very rapidly diffusing ions. It also may give a direct handle on the Bardeen–Herring correlation factor, particularly in the cases where the diffusing ion undergoes a very highly correlated motion through the lattice, as is expected for vacancy diffusion of impurities, where the binding energy between the impurity and the vacancy is very large. Our data indicate that the temperature dependence of the broadening of the Mössbauer line follows an Arrhenius relation, with a slope parallel to the classical diffusivity for cases which are not highly correlated. In the cases so far studied the broadening is only half of the earlier theoretically predicted value, and the significance of this point is discussed. A novel furnace assembly used for hot-source and hot-absorber experiments under 1100°C *is given.*

INTRODUCTION

In 1960 Singwi and Sjölander [1] predicted that the zero-phonon Mössbauer resonance should broaden in proportion to the diffusivity.

*Work supported in part by the Advanced Research Projects Agency.

†Present address: Sandia Laboratories, Organization 5131, Albuquerque, New Mexico.

Their model of diffusive motion was highly idealized. They assumed that there were oscillatory motions in a harmonic well followed by diffusive jumps or translations. They explicitly treated two limits of this model, the continuous-diffusion case and the sudden-jump approximation. The first corresponds to having the ion continuously in the translational state, and the second corresponds to having the fraction of time spent in translation negligibly small compared to the time spent in the oscillatory state. For the continuous diffusion model, which is analogous to liquid diffusion, Singwi and Sjölander predicted

$$\text{Case (i):} \qquad \Delta\epsilon = 2E_0^2 D/\hbar c^2$$

where $\Delta\epsilon$ is the broadening of the Mössbauer line, E_0 is the energy of the emitted γ-ray, and c is the velocity of light. As an alternative model, Singwi and Sjölander also considered the sudden-jump approximation, which gave the diffusion broadening as

$$\text{Case (ii):} \qquad \Delta\epsilon = 2\hbar(1 - \alpha)/\tau$$

where α is a term which can be shown to be negligible for $kr_0 \gg 1$, a condition that applies in every case considered here (k is the wave number of the emitted γ-ray, and r_0 is the nearest-neighbor distance, or the distance of one jump), and τ is inversely proportional to the diffusivity D. This second case is analogous to case (i), except the proportionality constant is different. For cubic solids the exact relation between τ and D is given by $D = r_0^2 f_c/6\tau$, where f_c is the Bardeen–Herring correlation factor, which is unity for a truly random-walk process. Diffusion in the systems considered here is believed to be *via* the vacancy mechanism, however, for which f_c is less than unity because the vacancy distribution is asymmetrical following an initial vacancy–ion exchange.

Although the correlation factor f_c is a constant of order unity for self-diffusion, it can be orders of magnitude smaller for cases of impurity diffusion, when there is a strong binding between the impurity and the vacancy. Thus, when the impurity–vacancy exchange rate is very rapid compared to the exchange rate for host atoms, it is essentially the host–vacancy exchange rate which limits the macroscopic diffusivity.

The theory of Singwi and Sjölander does not explicity include correlation effects, and hence assumes a purely random-walk process. We have assumed that this analysis will apply whether or not f_c is less than unity. Thus, we use their analysis with τ being the mean time that an ion is in a specified lattice site independent of correlation effects. This assumption should not lead to any serious difficulties for correlated diffusion, where $f_c \approx 1$, such as is observed for iron in copper or gold. The validity of this assumption would have to be examined more carefully for highly correlated diffusion, such as might be expected for tin diffusion in the noble metals. If this assumption is even approximately true for highly correlated systems of this latter type, then diffusion-broadening in conjunction with standard tracer-sectioning techniques could be used to measure directly the correlation factor f_c.

A convenient way of characterizing the broadening of the Mössbauer line is to write the ratio of the broadening to the natural linewidth Γ_N in the following way:

$$\Delta\epsilon/\Gamma_N = \sigma(k^2\tau_N)D$$

where σ is a dimensionless constant which is model-dependent. For the two models discussed we can write the broadening parameter σ as follows:

Case (i): $\sigma = 2$, for all examples.

Case (ii): $\sigma = 12/f_c r_0^2 k^2 \approx 0.04$ for Fe^{57} diffusion in metals.

Singwi and Sjölander's consideration of the above models as well as the intermediate cases suggest that we might expect to observe a value of σ approximately as indicated in (ii), but possibly between cases (i) and (ii).

The applicability of this technique is limited to a fraction of the available Mössbauer isotopes, although some of the isotopes having relatively broad lines may be more useful for this type of study than first supposed, if correlation effects are appreciable. Table I shows two Mössbauer isotopes where diffusion broadening might be observed at temperatures near the melting point.

The last column shows that for cases which are expected to be highly correlated, i.e., where $f_c \approx 0$, the ratio of the nuclear mean life τ_N to the mean life τ for making an atomic displacement may be very large, and broadening effects 100 times the natural linewidth are not unreasonable.

Despite the early prediction of diffusion broadening of the Mössbauer line, no direct measurements of the effect in solids had been made until the present investigations [7, 8] were carried out.

Table I. Estimated Diffusion Broadening near the Melting Temperature for Some Mössbauer Isotopes Diffusing in the Noble Metals.

Source/Host	Ref.	Diffusion constant at host melting temperature, cm^2/sec	f_c	τ_N/τ
Fe^{57}/Cu	[2, 3]	6×10^{-9}	0.8*	10
Fe^{57}/Ag	[2]	5×10^{-9}	0.8*	6
Fe^{57}/Au	[4]	1.2×10^{-8}	0.8	15
Sn^{119}/Cu	[5]	$\sim 6 \times 10^{-8}$	0.1†	~150
Sn^{119}/Ag	[6]	3×10^{-8}	0.1†	~60

*The isotope measurements of ref. [2] assure that these values are not in error by more than 20%.
†Only a rough estimate.

This is perhaps connected with the intrinsic difficulties associated with careful measurements of line shape at elevated temperatures.

In order to heat the samples under investigation to the temperature region where diffusion broadening was observable, we constructed a furnace which would satisfy the conditions of this experiment, as none was available commercially. The furnace had to be capable of reaching temperatures as high as 1060°C, be free from thermal gradients across the sample, and transmit low-energy gamma rays with a source-to-detector distance of less than 10 in.

The basic design we employed, shown in Fig. 1, consists of concentric quartz tubes, the inner tube used to hold the sample solidly, and the outer tube used to seal the sample from the atmosphere. The ends of the outer tube are sealed with O-rings to water-cooled caps, which serve to hold the assembly in place by compressing the inner tubes. Special high-temperature heater elements are placed around the tubes, with an Inconel block serving to distribute the heat uniformly over a region large compared to the sample size. Two methods were used to contain the sample inside the furnace. Illustrated in Fig. 1 and in more detail in Fig. 2 is the furnace assembly used for experiments in which the radioactive source was heated. When the absorber was heated the sample was sandwiched between two boron nitride disks and held in place by two iron rings, which were covered in front and back by thin boron nitride disks to provide an isothermal enclosure.

In most of the experiments we employed a hydrogen atmosphere inside the furnace to increase thermal transfer to the sample. Almost

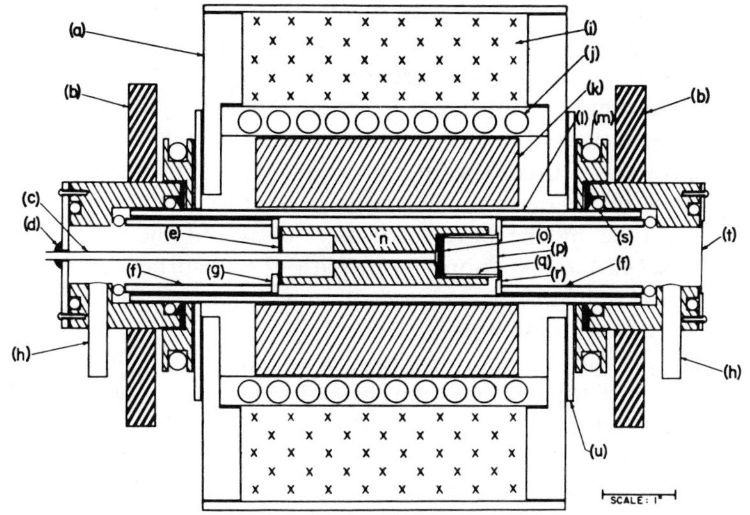

Fig. 1. Components and assembly of the furnace used for the case of a hot source: (*a*) transite cover, (*b*) end-cap clamps, (*c*) quartz protection tube for thermocouple, (*d*) Silastic RTV sealant, (*e*) rear boron nitride baffle, (*f*) inner quartz tubes, (*g*) stainless steel ring, (*h*) inlet and outlet tubing, (*i*) insulation, (*j*) heater elements, (*k*) Inconel cylinder, (*l*) outer quartz tube, (*m*) water-cooling assembly, (*n*) copper source holder, (*o*) boron nitride discs and sample assembly, (*p*) boron nitride baffle, (*q*) quartz compression tube, (*r*) stainless steel ring, (*s*) O-ring seal and compressor, (*t*) exit window assembly, (*u*) transite cover.

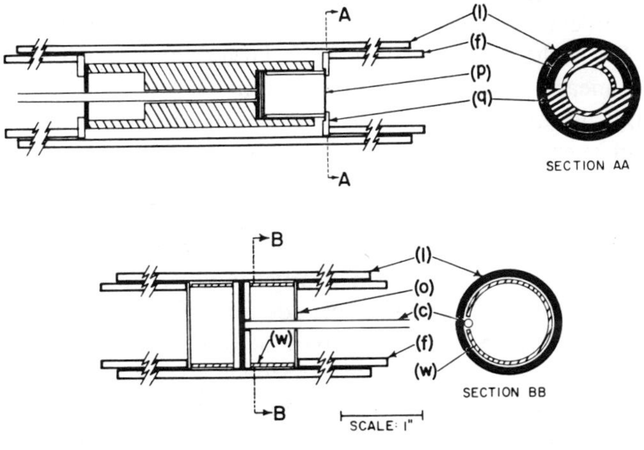

Fig. 2. Components of the core of the furnace assembly used for the case of a hot source (top) and a hot absorber (bottom). Letters labeling components are the same as in Fig. 1. The label *w* refers to the iron rings not shown in Fig. 1.

all thermocouple materials are severely affected by hydrogen at elevated temperatures, so it was found sufficient to provide protection in the form of a thin-walled quartz tube, vented to the outside. It was necessary, however, to replace the thermocouple every 12–24 hr at the higher temperatures, due to hydrogen seepage through the thin-walled quartz protection tube. The thermocouple (not shown in Fig. 1 or Fig. 2) was made of chromel–alumel. To prevent damage to the protection tube by the boron nitride, we covered the tube with a 2-mil (0.002-in.) thick copper foil. With the absorber holder, the quartz protection tube was placed in contact with the boron nitride disks which held the sample. In order to prevent the protection tube from masking the γ-ray transmission, it was placed along the side of the inner quartz tube, rather than through the center, as was done for the source holder. Direct measurements of thermal gradients were made in both the source and absorber holders. We determined that the greatest error in thermometry was less than 1°C.

For one set of experiments it was necessary to evacuate the inner chamber of the furnace to a pressure of the order of 10^{-6} Torr. Although thermal transfer to the sample holder from the outer quartz tube was not as good as with the hydrogen fill, it was possible to maintain a uniform temperature across the sample to within ± 1°C, using the same components as were used in the hydrogen furnace. Using commercial controllers, either of the time-proportioning or the current-proportioning type, it was possible to regulate the temperature of the sample to within ± 2°C when a hydrogen atmosphere was present. With the vacuum furnace a larger control dispersion of ± 4°C was measured, due to the larger thermal lag between the external heater and the sample holder. Thus, the maximum error in temperature for those experiments done in hydrogen was ± 3°C, and for those done under vacuum the error was ± 5°C.

Since vibrational motion can severely affect the Mössbauer spectrum, care was taken to mount all components rigidly. As a check, we ran known spectra with samples mounted in the furnace and found no evidence for vibration broadening of the Mössbauer line.

The boron nitride used to clamp the samples was made from high-purity material. After one series of experiments the discs were checked by running a spectrum without the sample present. It was found that the boron nitride was free from contaminants which would alter the spectrum and had not picked up any foreign substances from the samples in the course of the experiment.

If the furnace is moved to provide a Doppler shift, as was the case with the absorber experiments, care must be taken in the selection of end-window materials. To prevent the windows from acting as resonant absorbers, high-purity boron nitride was employed. A direct measurement, with the absorber removed, indicated negligible resonant absorption by these windows. For the source experiments, in which there is no relative motion of the furnace, we used a thin beryllium disk for the front exit window.

Although mass transport in solids, usually referred to as diffusion, has been measured by many techniques, the most common accurate procedure in recent years has been the tracer-sectioning technique. The procedure involves electroplating a thin layer of a radioisotope whose diffusion constant is desired, and letting this initial delta-function distribution disperse into the host crystal at a known temperature, usually within a few hundred degrees of the melting temperature. The diffusivity is then determined by slicing the specimen with a lathe or microtome and determining the diffusion profile. Within a given laboratory it has been possible to obtain a reproducibility in diffusion constants by this technique to within a few per cent. Classical diffusion measurements are difficult, however, if the diffusivity is either too great or too small. In the former case the warm-up correction (the correction to the actual time of carrying out the experiment resulting from the delay in heating the sample to the equilibrium value of the experiment) is prohibitive, and in the latter case the diffusion anneal time is too great for practical experiments. It is noteworthy that in the temperature region where the diffusivity is so great as to handicap tracer-sectioning experiments the Mössbauer approach is particularly well suited, as the problem of excessive penetration into the sample does not occur for this technique.

Unfortunately, relating macroscopic diffusion data to fundamental atomic-jump processes demands a very long list of assumptions, many of which cannot be justified in a simple and unambiguous way. For example, most theories begin by assuming a mechanism of diffusion, such as the vacancy mechanism, which requires that a tracer ion exchange with a vacancy in order to move through the lattice. Prior to the recent studies of the mass dependences of diffusion in solids [2], for which few measurements exist even to date, the vacancy mechanism was supported only by a theoretical calculation due to Huntington and Seitz [9] for the case of copper. The possibility of another mechanism, such as cracks in the crystal, was always dismissed by indirect arguments about the expected and experimental diffusion profiles, and claims of extreme linearity in the Arrhenius plots of ln D *versus* $1/T$. These arguments lack a convincing quality because of the limited temperature range over which most macroscopic diffusion measurements are made ($\sim 300°C$). The difficulties of these indirect arguments can be seen by noting that diffusion in many ionic crystals which was formerly thought to be *via* the vacancy mechanism is now believed to involve two mechanisms, i.e., vacancies and divacancies. Also, evaporation effects [3], such as are found for iron diffusion in copper, and solubility-limiting effects [2], such as occur for iron diffusion in silver, introduce some uncertainty into the meaning of the standard type of diffusion measurement, which can often be eliminated by the Mössbauer type of study. The Mössbauer effect gives a unique way of circumventing many of the problems inherent in the usual macroscopic measurements because this technique is microscopic and senses slight differences in environment through the various parameters obtained in the measured Mössbauer resonances. The technique has the added advantage that it permits reproducibility checks not possible with the tracer-sectioning technique, because the Mössbauer technique does not destroy the sample in the measuring process. This nondestructive feature of the Mössbauer approach also permits the measurement of the entire temperature dependence of diffusivity with a single sample of constant impurity content.

The broadening of the Mössbauer line for a case where the jump time is comparable to the nuclear lifetime is illustrated in Fig. 3 for the Fe^{57}-in-Au system. This is an illustrative example from among 16 such measurements on the iron-in-gold and iron-in-copper systems.

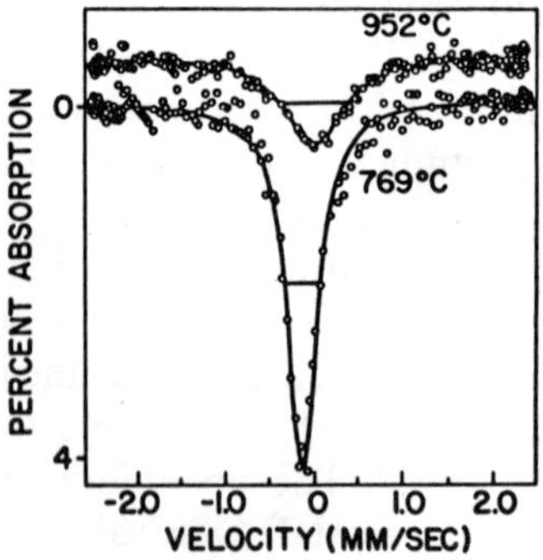

Fig. 3. Mössbauer spectra for a Co^{57}-in-Au source *versus* a $Na_4Fe(CN)_6 \cdot 10\ H_2O$ at 952°C and 769°C. The total number of counts per point in the upper curve is 10^6, giving an average statistical error per point of 0.1%.

It should be noticed that the width suddenly increases, with a corresponding dramatic decrease in the percentage dip, even though the temperature change is only from 769°C to 952°C, a change of less than 20% on the absolute temperature scale. Within the experimental errors the line shape is Lorentzian, as required by the sudden-jump theory.

Since the theory shows that the increase in the width of the Mössbauer line $\Delta\epsilon$ is proportional to D, and since it is fairly generally true that diffusion in solids follows an Arrhenius relation, $D = D_0 e^{-Q/kT}$, we would expect a plot of $\ln(\Delta\epsilon)$ as a function of $1/T$ to be linear with a slope equal to Q/k. Such a plot is given in Fig. 4 for the systems iron in copper and in gold. Indeed, to within the experimental errors the broadening is proportional to the measured microscopic diffusivity D.

It should be noted that because the parent nucleus for many Mössbauer nuclides, such as Co^{57}, has a different valence from the daughter isotope, there will generally be different electron screening, and hence different interactions with lattice vacancies. If the vacancy distribution for the parent does not relax to the equilibrium of the

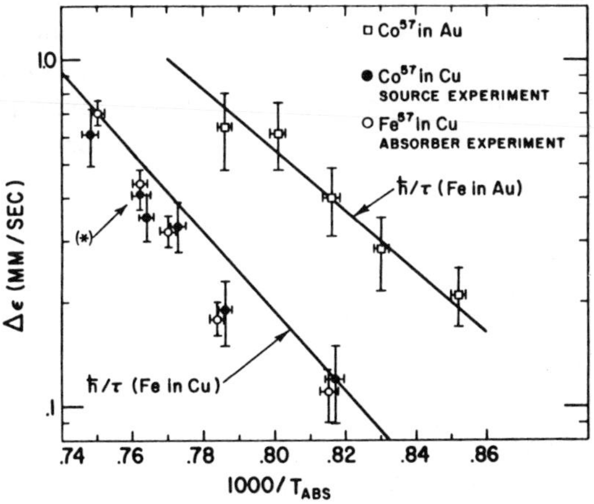

Fig. 4. Line broadening as a function of temperature. The vertical error bars reflect the dispersion in $\Delta\epsilon$ found from a least-squares fit of the data to a Lorentzian function. The errors in temperature indicate an upper bound of ± 3°K. The upper line and data points are for a Co^{57}-in-Au source *versus* a $Na_4Fe(CN)_6 \cdot 10\, H_2O$ absorber. The lower line gives the results for iron diffusion on copper. This earlier work shows that a Co^{57}:Cu source has the same temperature dependence to the broadening as a 1% Fe^{57}-doped Cu absorber. The point marked by (*) was done under vacuum with a vacuum-annealed source, while all of the other experiments were done under 1 atm of H_2 gas.

daughter in the lifetime of the isomeric transition, then the diffusion rate may not reflect that of the pure daughter isotope. Figure 4 shows, however, that essentially the same broadening occurs for both the hot-source and hot-absorber experiments, and that the new vacancy equilibrium for Fe^{57} diffusing in copper is achieved within the isomer lifetime of 10^{-7} sec.

Unfortunately, even though the line shape is approximately Lorentzian and the broadening of the Mössbauer line appears proportional to the diffusivity, there is a discrepancy between our results and the sudden-jump model treated by Singwi and Sjölander. The proportionality constant $\Delta\epsilon/D$ which we measure is very nearly 1/2 of the value predicted by theory. In terms of the broadening parameter σ, our observations indicate that $\sigma \approx 0.02$, only half of the value predicted in the sudden-jump model and two orders of magnitude less than required by the continuous-diffusion model.

There is an added experimental result which may be relevant to this discrepancy: in the region where diffusion broadening occurs we observed an apparently anomalous falloff in the recoil-free fraction f of emitted γ-rays, as illustrated in Fig. 5. The low-temperature part of the curve is based on the data of Steyert and Taylor [10], and the high-temperature portion is based on our results, using the area under the Mössbauer resonance as a measure of the recoil-free fraction. To explain this sudden falloff in f in terms of anharmonic effects requires more than the lowest-order anharmonic correction suggested

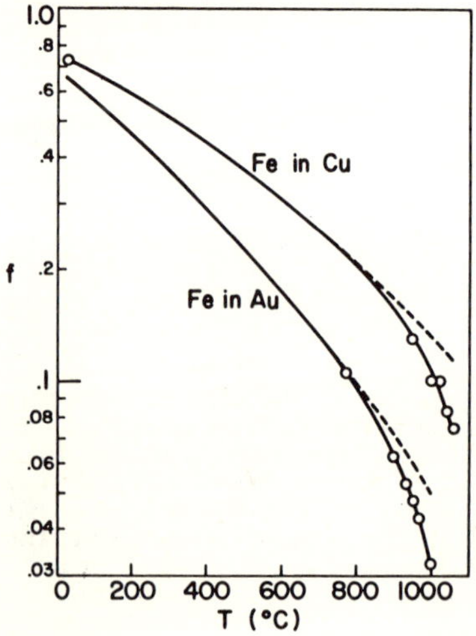

Fig. 5. Temperature dependence of the recoil-free fraction f. The dashed curve at elevated temperatures is an extrapolation of Steyert and Taylor's data including the first-order anharmonic correction which they give.

by Steyert and Taylor if we accept their value of the lowest-order anharmonic correction ϵ. It is possible that this effect is related to the diffusion-broadening phenomenon which occurs in approximately the same temperature region. To predict such effects, however, would require a more elaborate theory than has thus far been proposed. The Singwi–Sjölander calculation would indicate the usual Debye–Waller falloff for f in the region where diffusion broadening occurs. The extension of the theory to include effects of this kind, where the self-correlation function $G_s(\mathbf{r}, t)$ is non-Gaussian, is very involved, although the concepts are discussed by Rahman *et al.* [11]. These effects should alter the line shape from the Lorentzian as well as altering the full-width at half-maximum. Deviations from the Lorentzian shape were not discernible in the present experiments.

An alternative explanation can be given, namely, that the underlying assumption of the sudden-jump model, that an atom relaxes out of a lattice site with a time dependence of $\exp(-t/\tau)$, is not valid for the present case. Thus, the jump probability involves two factors: (1) the probability that a vacancy is next to the ion in question, and (2) the probability of an exchange with the vacancy when it is adjoining the ion. If the time dependence of relaxation were then more nearly a step function than an exponential, the expected broadening would be to first order $\hbar/\tau$, instead of $2\hbar/\tau$, in analogy with the result obtained for the classical time-filtering experiment [12]. The difficulty with this explanation is that we would expect a noticeable deviation from a Lorentzian line shape, which was not discernible in the present experiment. This point is not easily resolved experimentally, however, and to obtain the present experimental accuracy required approximately 10^6 counts per channel because the percentage effect at the highest temperatures is only of order 1%. Thus, a more-refined measurement of the linewidth and, in particular, the line shape will require a much greater effort than we have thus far attempted.

Considerable effort has been made in studying impurity diffusion in recent years to examine the effects of electron screening on the vacancy–impurity interaction. The results for cases such as tin diffusion in silver appear to be well-explained by the screening model, which predicts a significant binding energy between a tin ion and a host vacancy. This binding should lead to highly correlated jump sequence. Such correlated diffusion may afford favorable cases for studying diffusion by the Mössbauer technique, as the broadening should depend inversely on f_c. To see if we could observe these effects for Sn^{119} in Ag, we carried out a similar experiment to the Fe-in-Au and Fe-in-Cu cases. Unfortunately, we observed [13] a very rapid reduction in $f(T)$ above 300°C, and we thus far have only extended the experiment to 470°C, where no diffusion-broadening effects would be expected, and, indeed, where none were observed.

REFERENCES

1. K. S. Singwi and A. Sjölander, *Phys. Rev.* **120**:1093 (1960).
2. J. G. Mullen, *Phys. Rev.* **121**:1649 (1961).
3. C. A. Mackliet, *Phys. Rev.* **109**:1964 (1958).
4. D. Duhl, K. Hirano, and M. Cohen, *Acta Met.* **11**:1 (1963).
5. C. J. Smithells, *Metals Reference Book* (Butterworth and Co., London, 1962), Vol. II, 3rd ed.
6. C. T. Tomizuka and L. O. Slifkin, *Phys. Rev.* **96**:610 (1954).
7. R. C. Knauer and J. G. Mullen, *Phys. Rev.* **174**:711 (1968).
8. R. C. Knauer and J. G. Mullen, *Appl. Phys. Letters* **13**:150 (1968).
9. H. B. Huntington and F. Seitz, *Phys. Rev.* **61**:315 (1942); **76**:1728 (1949).
10. W. A. Steyert and R. D. Taylor, *Phys. Rev.* **134**:A716 (1964).
11. A. Rahman, K. S. Singwi, A. Sjölander, *Phys. Rev.* **126**:986 (1962); **126**:997 (1962).
12. F. J. Lynch, R. E. Holland, and M. Hamermesh, *Phys. Rev.* **120**:513 (1960).
13. R. C. Knauer, Ph.D. Thesis, Purdue University, unpublished.

Mössbauer Spectrometry as an Instrumental Technique for Determinative Mineralogy*

C. L. Herzenberg

IIT Research Institute
Chicago, Illinois

As a result of current availability of Mössbauer spectra for a large number of minerals, including examples from many of the important groups of rock-forming minerals, it is at present feasible in many cases to examine and interpret the Mössbauer spectra of rocks, sediments, soils, and other multiphase natural mineral aggregates. We have measured Mössbauer spectra for a considerable number of specimens of igneous rocks and in addition for samples of marine sediments, and a selection of such spectra are presented. Comparisons are made between the preliminary results of Mössbauer spectrometry on standard silicate rocks and data from chemical analyses and modal information from petrographic examinations. Further topics include total iron determinations, ferric-to-ferrous ratio measurements, improvement of petrographic modal information using Mössbauer spectrometry, identification of igneous rock types on the basis of Mössbauer spectral patterns, detection and identification of coarse-grained magnetite and hematite, with applications in the area of paleomagnetism, and the characteristics of Mössbauer spectra of marine sediments.

INTRODUCTION

Among the elements of which the Earth is composed, iron is one of the most abundant. It constitutes about 5% by weight of the continental crust and about 7% of the entire crust. Furthermore, it is so widely distributed in nature that few samples of the solid materials of the Earth are substantially without it. Accordingly, the natural solid

*Research sponsored in part by NASA.

inorganic materials of the lithosphere, such as rocks, soils, sediments, muds, and so on, typically contain this element in amounts suitable for investigation by Mössbauer spectrometry. All of these, of course, are heterogeneous aggregates of minerals in varying proportions, and use of Mössbauer spectrometry permits one to look selectively at the iron-containing mineral phases in these materials.

Investigation of the state of the iron in natural materials is valuable for a variety of reasons. For one thing, iron is exceptional among the other principal elements of the crust in that it exists commonly in two valence forms and passes relatively easily from one to the other as the result of oxidation or reduction, so that the valence state of the iron in a rock or sediment sample can provide special information on the geochemical history of the sample. Use of Mössbauer spectrometry of course permits one to obtain information on the distribution of the iron in the different oxidation states.

Another example is the expansion of interest in paleomagnetism during the last several years, which has focused attention on the solid-state physics and chemistry of rocks, and, in particular, on the properties of the rock components that are chiefly responsible for the magnetic properties. As it happens, essentially all of the minerals which account for the magnetic properties of rocks are iron-bearing minerals. An understanding of the mineralogy of rock magnetism requires, among other things, identification of the magnetic mineral phases, and in some cases grain-size information for these magnetic phases as well. We have found that under many circumstances the Mössbauer spectra of these magnetically ordered iron compounds permit easy detection and identification of the magnetic phases as well as sometimes providing a useful method of grain-size estimation for ultrafine-grained mineral phases.

Generally speaking, some of the important rock-forming minerals contain iron, and from the Mössbauer spectra one can learn about the distribution of iron between mineral phases. In a large proportion of the specimens which we have investigated we have found that this technique can lead to superior petrographic information, even in the case of specimens which have already been relatively extensively studied using most of the previously available standard techniques.

One of the points here is that certain petrographic information on a rock specimen (in particular, the modes), or the corresponding information on the composition of a mud or clay sample, can be a matter of some importance, and it is often rather arduous to obtain such data using standard techniques, especially for fine-grained rocks or poorly crystallized sedimentary material. Rocks consisting in whole or in part of clay minerals or clay-sized particles present special problems in petrography. In some cases, like some clays we have investigated (and perhaps also desert varnish), one may have mineral coatings on the particles which are not concentrated enough to pick up by ordinary X-ray diffraction techniques. In sediments the particle-size distribution can include material which is quite small, and the particles themselves may be aggregates of extremely finely divided material. Detection and identification of many of the opaque or ore minerals in petrographic analysis can be done much more simply in many cases by Mössbauer spectrometry than by previous techniques, and, as I mentioned, this is especially significant for the magnetic minerals.

PROCEDURES AND TECHNIQUES

Although, the idea of the development of Mössbauer spectrometry as an instrumental technique for the study and analysis of rocks, soils, sediments, and other specimens of the lithosphere is quite an obvious one, and is one which I am sure has occurred to many other researchers, curiously little has been done along these lines. There are a variety of reasons for this; one contributing reason is that while the specimen of rock (or sediment or soil) will be solid or solid powder, which makes its use natural for Mössbauer spectrometry, the several per cent of iron in it will in general be distributed through a number of different mineral phases, with possibly only a fraction of 1 % in any one phase. And, of course, only 2.19 % of the iron nuclei in a natural sample are the resonant isotope, Fe^{57}. Consequently, one will frequently be dealing with small individual absorption contributions, and, as a result, relatively large numbers of total counts are necessary to obtain a usable signal above the statistics. In the last several years the situation has greatly improved with the availability of more-intense Co^{57} sources and counting equipment suitable for higher-speed data acquisition. Using a 50-mCi Co^{57} source from New England Nuclear and an Austin Science Associates spectrometer, we are able to take useful spectra on rock specimens containing less than 1 % by weight total iron, and in individual phases we can detect and identify peaks of intensity equivalent to less than 0.1 % iron on a thoroughly routine basis. I think this can be extended to below 100 ppm with really only straightforward improvements on present equipment and techniques.

One further development which has contributed to making this application of Mössbauer spectrometry feasible today is the current availability of Mössbauer spectra for quite a large number of minerals, including examples from many of the important groups of rock forming minerals. While there have been some recent publications containing useful but limited collections of Mössbauer data on minerals and mineral systems [1–4], most of this information is still scattered through the literature or unpublished.

I am not going to cover the application of Mössbauer spectrometry to the study of minerals *per se*, although a good deal of interesting work has been done here. This application area was recently reviewed very nicely by Hafner and Ghose [2, 3] in a lecture series under the auspices of the American Geological Institute.

In passing, however, I would like to mention that this type of measurement has been of some interest in connection with structural investigations of crystals [1–4]. Mössbauer spectrometry has also been shown to provide a useful technique for obtaining information on the previous thermal and pressure history of certain minerals, and therefore their host rocks, from the relative population of different lattice sites [5, 6]; the relative amount of iron in certain coexisting mineral phases in igneous and metamorphic rocks may provide similar information [7]. Mössbauer spectrometry has also been useful as an adjunct technique in the identification of new minerals [7].

Application of Mössbauer spectrometry to the quantitative analysis of multiphase samples has been dealt with in a number of earlier publications [8–14], and the general procedures which have proved most useful have been treated in varying detail in these publications as well as others. We have made no essential innovations in spectral-analysis procedures, so I will not discuss at length this aspect of our work.

Our recognition of the separate contributions in a complex spectrum comes from a comparison with the spectra of samples prepared from individual mineral phases, or from previously measured spectra or spectral parameters which have been made available to us from other laboratories,* or which have been obtained from the literature.

*We wish to thank, in particular, the following individuals for making available to us certain unpublished information: P. Imbert, S. Hafner, L. H. Bowen, S. K. Banerjee, A. v. d. Giessen, A. H. Muir, Jr. and G. M. Bancroft.

Our estimates of the relative amount of iron in the various mineral phases present in a sample are based on the usual procedures of area analysis for phase-concentration determination [8–14]. A great deal still needs to be done to improve these procedures for this particular type of application. For one thing, there are still no suitable published measurements of the resonant fractions for almost all of the minerals which we have to deal with. Also, in these measurements adequate corrections have still to be made for intensity asymmetries, nonresonant absorption differences, thickness corrections, and other associated effects. I should perhaps add that while most of the work I am reporting has been done with room-temperature measurements, in certain cases, particularly for many sediments, it is highly desirable to obtain spectra at reduced temperatures also (largely for the analysis of superparamagnetic material, and also to assist in the identification of particular phases), and even for many of the most-common and important minerals no spectra have yet been reported at other than room temperature.

Our measurements of the relative proportion of iron in different oxidation states have depended upon area-analysis techniques coupled with identification of the particular phases responsible for the different absorption contributions. The corresponding ferric-to-ferrous ratios that we have obtained have shown a rather discouraging lack of reliable quantitative agreement with ferric-to-ferrous ratios determined by the usual chemical techniques, although the qualitative behavior follows the expected trend, and in many cases we get reasonable quantitative agreement. The greatest discrepancies have usually occurred for cases in which chemical analyses corresponded to either especially high or especially low ferric-to-ferrous ratios; this has also been the experience of other groups [12]. We think we understand some of the major contributing factors that account for discrepancies in the ferric-to-ferrous ratios, and that considerably improved values will be possible.

In the determination of total iron, we have generally relied on area analysis in conjunction with an accompanying control measurement of an additional absorber of known total iron content—usually a geochemical standard of somewhat similar composition to the specimen under investigation. Our total iron estimates have in most cases turned out to be within about 15% of the correct values, although occasionally the discrepancies have been larger. Higher accuracies for total iron, ferric-to-ferrous ratio, and relative phase concentration have been reported in controlled experiments on simple mixtures of compounds [12], which supports our belief that considerable improvements will be possible in our quantitative results.

I would like to emphasize that we consider this preliminary work: we have neither the excellent statistics and quality of spectra nor, more particularly, the type of data analysis that we want yet. But we think these preliminary survey results are indicative of the great potential value of the Mössbauer technique in this application and that these results are interesting enough to warrant a description at this early stage of the work.

ROCK SPECTRA

The measurements that I am going to describe have been done by Dennis Riley of IIT Research Institute and myself. Our program of measurements of the Mössbauer absorption spectra of rock specimens so far has emphasized igneous rocks.

Our initial studies to ascertain the general character of rock spectra were made on various rock types from a number of sources.* One point that we found out fairly early is that the general petrologic character of an igneous rock can often be recognized rapidly from its Mössbauer spectrum. The spectra show quite characteristic and easily recognizable changes over the range of rock types between ultrabasic and acidic, as can be seen from the various figures discussed below, which have been selected from our measurements on the US Geological Survey standard silicates [15, 16].

Figure 1 is a dunite, an ultrabasic rock (this particular sample is DTS-1). Dunites are practically monomineralic rocks, containing nearly all olivine, largely magnesian olivine. The two strong lines in this spectrum are from the olivine, a ferromagnesian silicate mineral $(Mg,Fe)_2SiO_4$, which is responsible for almost all the resonant absorption in this sample. Petrographic information on this sample indicates that it contains over 99% olivine [15]. There is 6.19% by weight total iron in this sample. (It should be noted that this and subsequent figures are plotted from raw uncorrected data, and instrumental baseline dips are present in the spectra, in addition to the true

*Specimens were made available to us through the courtesy of Dr. Bert Woodland of the Field Museum of Natural History in Chicago; Francis J. Flanagan of the US Geological Survey; and Thomas C. Atchison and Rolland L. Blake of the US Bureau of Mines.

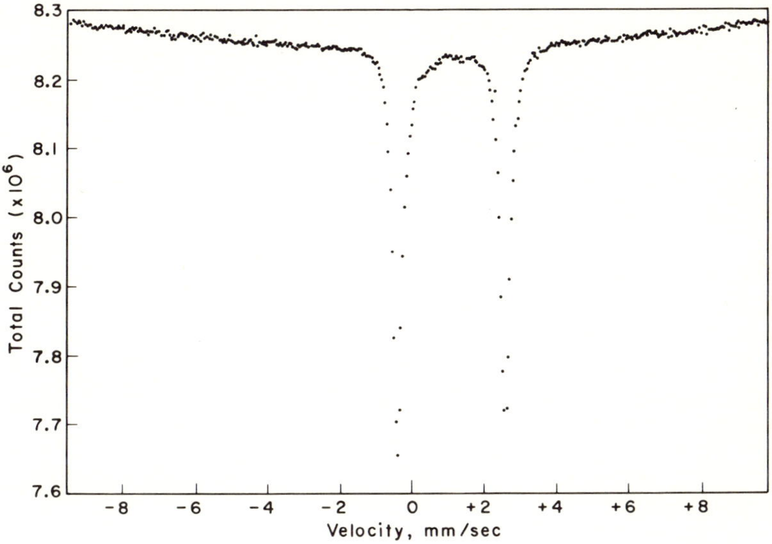

Fig. 1. Mössbauer hyperfine absorption spectrum of USGS dunite DTS-1.

absorption. All velocities are indicated with respect to the center of the iron spectrum.)

Figure 2 is a peridotite (PCC-1) containing 5.97% iron. This is a coarse-grained ultrabasic rock, rich in olivine, but containing, in addition to olivine, other mafic minerals such as orthopyroxene, as can be seen from the additional absorption. Our total iron measurement gave 5.6%, in fair agreement with the analytical chemistry. Detailed comparisons of our total iron and ferric-to-ferrous ratio determinations, as well as our results on the distribution of iron in the different mineral phases for these rocks, are given in Tables I and II.

In Fig. 3 we move away from the ultrabasic rocks to look at a basic rock. The character of the spectrum is changing. This figure is the interior portion of the absorption spectrum of a basalt, a common type of volcanic rock. Basalts have fine-grained texture and contain appreciable amounts of plagioclase as well as the ferromagnesian minerals. This specimen (BCR-1) has 9.45% total iron [15].

Figure 4 is a somewhat less-basic rock, an andesite (AGV-1). The olivine is mainly gone now, as can be seen from the velocities at which the main absorption is occurring. Andesite is also a fine-grained volcanic rock. This particular sample contains 4.78% total iron. Among the interesting features of this spectrum are the broadened

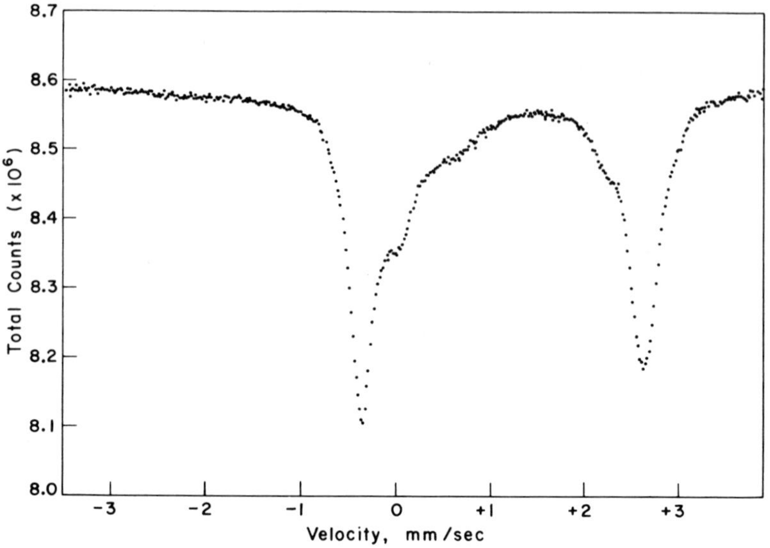

Fig. 2. Mössbauer hyperfine absorption spectrum of USGS peridotite PCC-1.

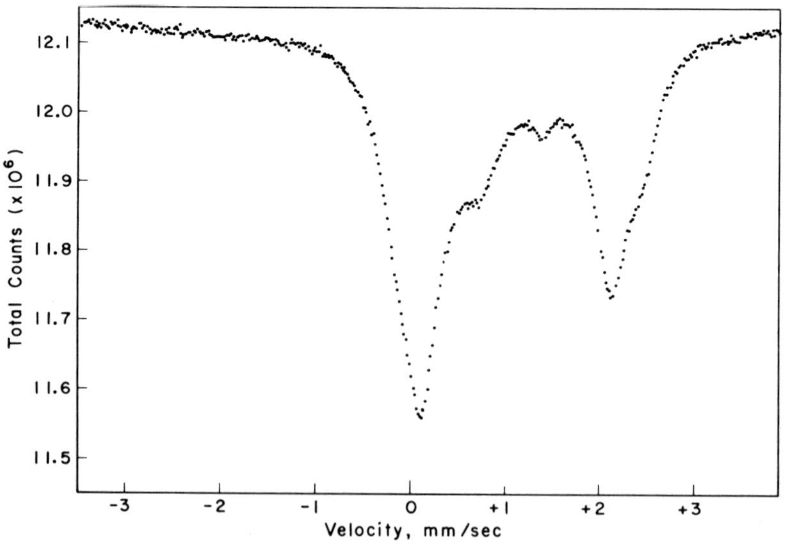

Fig. 3. Mössbauer hyperfine absorption spectrum of USGS basalt BCR-1.

Table I. Preliminary Total Iron and Ferric-to-Ferrous Ratios Determined for USGS Silicate Rock Standards

Sample designation and rock type	Total iron, % (chemical analysis)	Total iron, % (Mössbauer analysis)	Fe^{3+}/Fe^{2+} (chemical analysis)	Fe^{3+}/Fe^{2+} (Mössbauer analysis)
G-2 granite	1.94	1.9	0.67	0.71
GSP-1 granodiorite	3.03	2.9	0.69	0.62
AGV-1 andesite	4.78	4.0	1.98	1.91
PCC-1 peridotite	5.97	5.6	0.52	0.22
DTS-1 dunite	6.19	(6.19)	0.16	0.03
BCR-1 basalt	9.45	9.3*	0.37	0.36†

*Of this value, 6.4% is contributed by the pronounced low-velocity absorption spectrum. Correction for additional absorption present at high velocities as very diffuse peaks brings the value to 9.3%, but this additional absorption is more difficult to determine with accuracy.

†Based on tentative identification of the oxide as a titanomagnetite of composition approximately 40% magnetite.

Table II. Distribution of Iron in Major Identified Mineral Phases in USGS Silicate Rock Standards by Mössbauer Spectrometry (Preliminary Results)[a]

Sample designation and rock type	Mineral phase: Olivine	Pyroxene (Fe^{2+})	Biotite (Fe^{2+})	Ilmenite	Chromite	Bulk iron oxides[b]	Additional ferric iron
G-2 granite	—	—	44[c]	6	—	26[d]	24[e,f]
GSP-1 granodiorite	—	—	42	14	—	27[g]	17[e,f]
AGV-1 andesite	—	25	—	9	—	25[h]	41[f,i]
PCC-1 peridotite	70	8	—	—	4	—	18[j]
DTS-1 dunite	94	2	—	—	1	—	3
BCR-1 basalt	4	39[k]	—	8	—	32[l]	16[f,m]

[a]The percentage of total iron in each phase is tabulated.
[b]This does not include extremely fine-grained (superparamagnetic) magnetite or hematite, which do not exhibit appreciable Zeeman interaction. High-titanium-content titanium-iron oxides are also excluded.
[c]This may include a contribution from a small amount of chlorite present in sample G-2.
[d]Largely magnetite.
[e]This is presumably largely ferric iron in the octahedral layer of the biotite.
[f]Some contribution to the additional ferric iron may come from very limited amounts of iron in the feldspars.
[g]Some hematite appears to be present in addition to the magnetite.
[h]The oxide spectrum shows highly broadened peaks located at energies corresponding to an internal field reduced below that of bulk hematite. The character of the spectrum is similar to what has been observed for solid solutions in the $FeTiO_3$–α-Fe_2O_3 system, and the observed internal field corresponds to an ilmenohematite of about 70% hematite composition, or this might possibly be ferric oxide dispersed in the glass present in the andesite. Maghemite, magnetite, and finely divided superparamagnetic hematite also have peaks in these vicinities.
[i]Some contribution to the additional ferric iron is expected to come from ferric iron in the iron-bearing glasses present in this sample and additionally from extremely fine-grained superparamagnetic hematite.
[j]This may originate from ferric ions in the serpentine and may also include a contribution from ferric ions in the pyroxene.
[k]This entry includes, in addition to the contribution from the augitic pyroxene, a further ferrous contribution which may be from ferrous iron in the iron-bearing glasses present in this basalt. In addition, another ferrous component is present having parameters similar to an amphibole or a trioctahedral mica.
[l]This basalt exhibits a diffuse absorption by unresolved peaks at higher velocities beyond the obvious inner absorption peaks. We tentatively ascribe this broad, shallow absorption to the presence of titanomagnetites, which give very broad magnetic spectra.
[m]This presumably comes in part from ferric iron in the augite as well as ferric iron in the iron-bearing glasses.

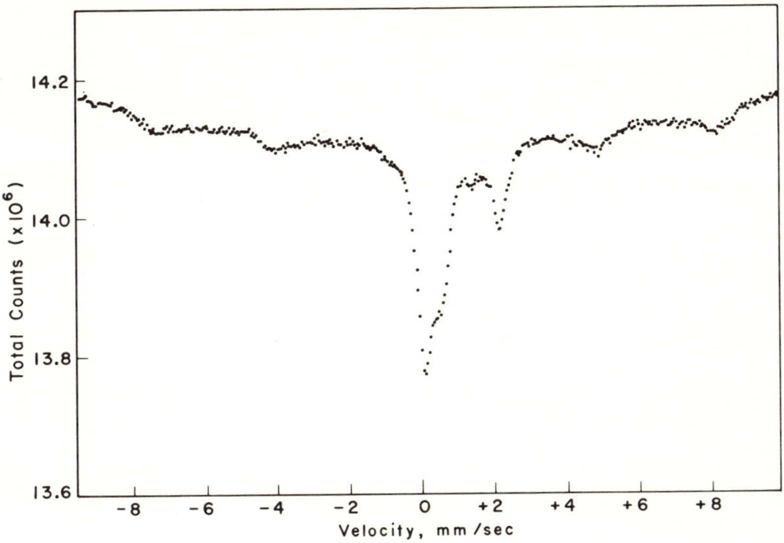

Fig. 4. Mössbauer hyperfine absorption spectrum of USGS andesite AGV-1.

oxide peaks at high velocities; we think this may be either a titanohematite, or else it may originate from iron oxide dispersed in the glassy portion of the groundmass of this sample. Our ferric-to-ferrous ratio determination for this sample was 1.91, while 1.98 was the corresponding value obtained from chemical analysis [15].

In Fig. 5 we now move further toward the acidic end of the classification of igneous rocks; this is a granodiorite (GSP-1). Granodiorites are coarse-grained rocks containing appreciable quartz, and roughly equal amounts of orthoclase and plagioclase. Our determination of 2.9% total iron for this sample is in reasonable agreement with the 3.03% total iron content determined by chemical techniques [15]. The absorption is due largely to iron in the biotite mica phase, with oxide peaks also evident at the higher velocities.

The last figure in this sequence, Fig. 6, is a granite, which is the most-acidic (silicic) rock in this sequence. This is a coarse-grained plutonic rock, that is, one that has solidified at depth rather than in a surface environment. This sample has a total iron content of 1.94% by weight, distributed through several phases; these can be identified as mainly biotite, and the oxide which gives the absorption in the wings. Our total iron determination in this case agreed with the

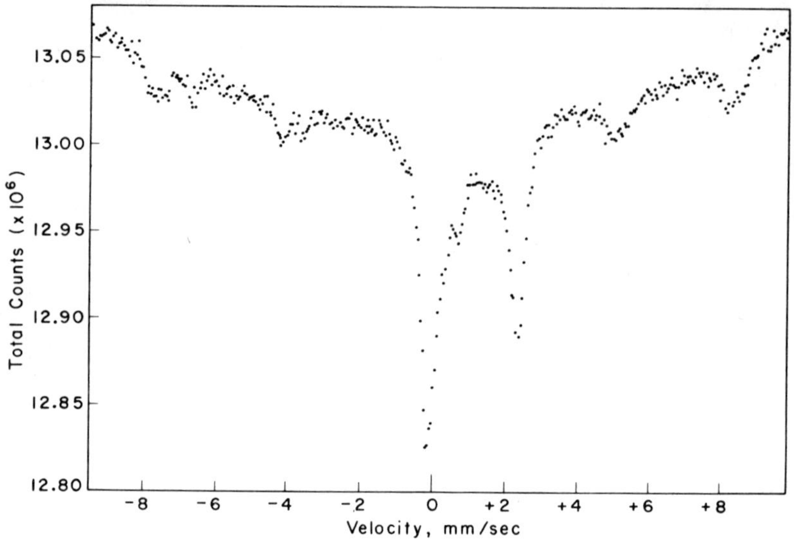

Fig. 5. Mössbauer hyperfine absorption spectrum of USGS granodiorite GSP-1.

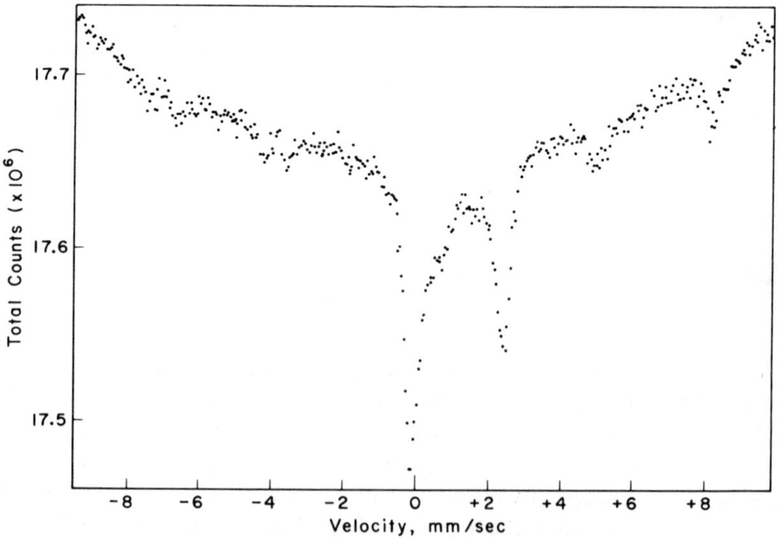

Fig. 6. Mössbauer hyperfine absorption spectrum of USGS granite G-2.

established value, and our estimate of the ferric-to-ferrous ratio was fairly close, 0.71 *versus* the 0.67 chemical value.

Earlier I mentioned the importance of the oxide minerals, as most of the magnetic minerals which play significant roles in paleomagnetism are oxides or related minerals.

Magnetite shows up very prominently in Fig. 7, which is a rock standard prepared by the Bureau of Mines. This is a serpentinite [17, 18], which is an intrusive igneous rock. Serpentinites, as their name implies, are composed largely of minerals of the serpentine group, all of which are usually ascribed to late- or post-magmatic hydrothermal alteration of pyroxene and olivine in pyroxonite, peridotite, and dunite. Interior peaks are present which are ascribable to absorption by mineral phases including olivine, augite (which is a monoclinic pyroxene), chlorite, and chromite in the specimen. Our determination of total iron in this case was 5.79 %, *versus* reported 5.88 %. However, we found much less ferric iron than was reported in the chemical analysis; our estimate of the ferric-to-ferrous ratio was 1.4, *versus* a reported 3.2.

The next spectrum (Fig. 8) shows not only the presence of magnetite, but also the presence of a great deal of ilmenite, probably

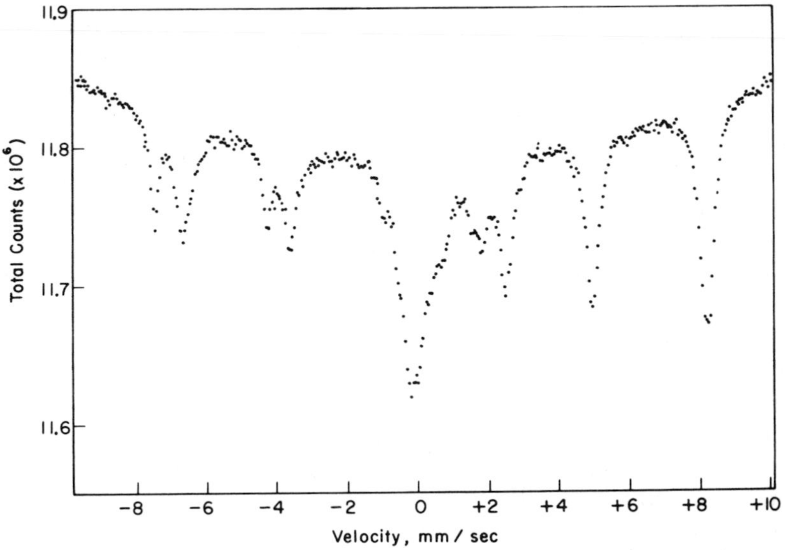

Fig. 7. Mössbauer hyperfine absorption spectrum of Bureau of Mines serpentinite SLR 1B-9.

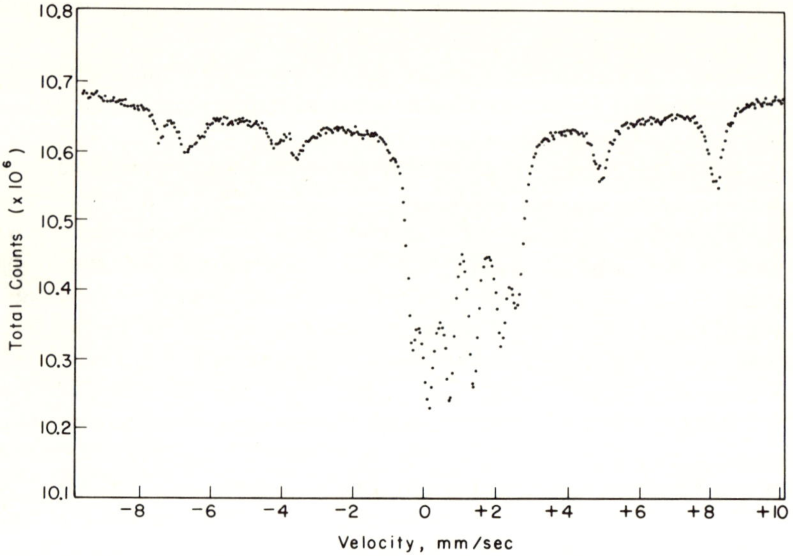

Fig. 8. Mössbauer hyperfine absorption spectrum of Bureau of Mines gabbro SLR 1B-7.

exsolved ilmenite. This is a gabbro, a coarse-grained basic rock, the intrusive equivalent of a basalt [17, 18]. The two innermost peaks in this spectrum, near 0.7 and 1.4 mm/sec, are the ilmenite contribution; the outer four peaks in the paramagnetic region originate from olivine and pyroxene in the sample; these may be ulvöspinel absorption also. Our estimate of total iron in this rock specimen gave 13.4%, a little higher than the reported value of 12.28%. In this case also we got a very serious discrepancy between the ferric-to-ferrous ratio estimate of 0.15 from the simplest interpretation of the spectrum, *versus* 0.58 based on the chemical data.

These last two cases have been of rocks containing magnetite (more precisely, titanomagnetite), having peaks that are relatively easily recognizable, although somewhat broadened, with possibly some spurious features. However, the oxide minerals in some rocks—many basalts, for example—frequently have a high titanium content, and one has to cope with titanomagnetites of intermediate composition. These have been shown [19–22] to exhibit spectra which can be greatly broadened, and for some naturally occurring compositions one gets very wide, practically featureless absorption troughs.

Figure 9 provides an example of this, although it is somewhat difficult to distinguish the broad diffuse absorption visually because this figure is raw data which has not been corrected for baseline curvature. This is a tholeiitic flow basalt from the Bureau of Mines [17, 18]. Data analysis on spectra of rocks containing small amounts of titanium-iron oxides of intermediate composition is still in a very unsatisfactory state. In addition to the broad absorption trough from the titaniferous magnetite (which we think corresponds to about a 60–70 % ulvöspinel composition in solid solution), there also visible are small ilmenite peaks in the interior of the paramagnetic portion of the spectrum. Lines from olivine and chlorite, and augite, ulvöspinel, and glass are prominent in this spectrum.

Figure 10 is the spectrum of an obsidian, a natural glass [17, 18]. This contains 1.55 % total iron by weight which can be compared with our determination of 1.65 %. The ferric-to-ferrous ratio is 0.32, in good agreement with our estimate of 0.29 made on the basis of the Mössbauer area analysis. This spectrum shows mainly broadened peaks, and is not particularly structured, as happens with some glasses. Since studies of synthetic glasses have shown composition-dependent spectral shapes [11, 23–26], we think that more measurements of obsidians may be of interest, and perhaps this technique

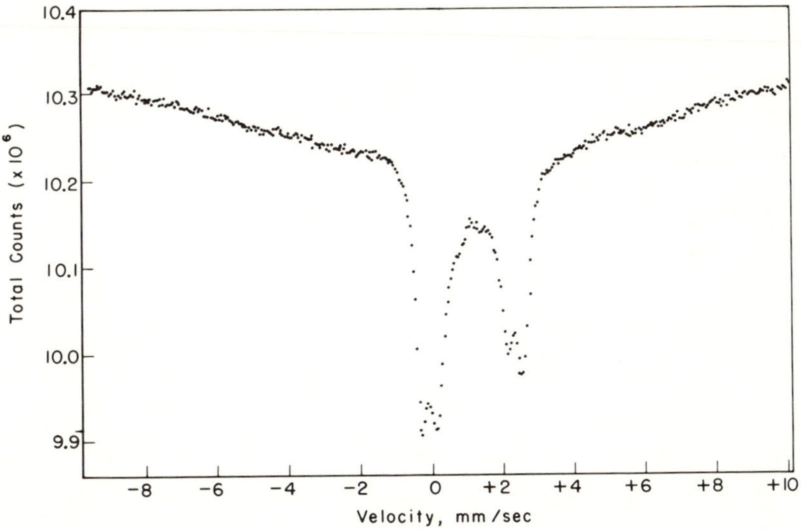

Fig. 9. Mössbauer hyperfine absorption spectrum of Bureau of Mines flow basalt SLR 1B-1.

might even make a contribution to the tracing of the origins of obsidian specimens of importance in archaeology, as neutron-activation and trace-element studies have done recently.

In Fig. 11 one can again see the natural glass contribution, with some additional absorption also, partly originating from crystalline material. This is the spectrum of a dacite, an extrusive rock of composition equivalent to quartz diorite, intermediate between rhyolite and basalt. It has a glassy groundmass with microphenocrysts or inclusions of crystalline materials in the rock. Also visible is evidence for broadened oxide peaks at higher velocities: magnetite, in this case. Our total iron estimate for this dacite specimen was 2.13%, compared to the reported 1.92%. The excursion in the ferric-to-ferrous ratio determination was again large (0.50 Mössbauer *versus* 0.29 chemistry), which I think is attributable in part to a poor oxide determination. Incidentally, on the basis of the spectral contribution of iron-bearing glasses, when these are present, we are often able to distinguish between extrusive rocks and intrusive or plutonic types, an additional interesting feature for instrumented petrography.

SEDIMENT SPECTRA

Next I will discuss some of the Mössbauer spectra we have taken for various types of marine and freshwater sediments.

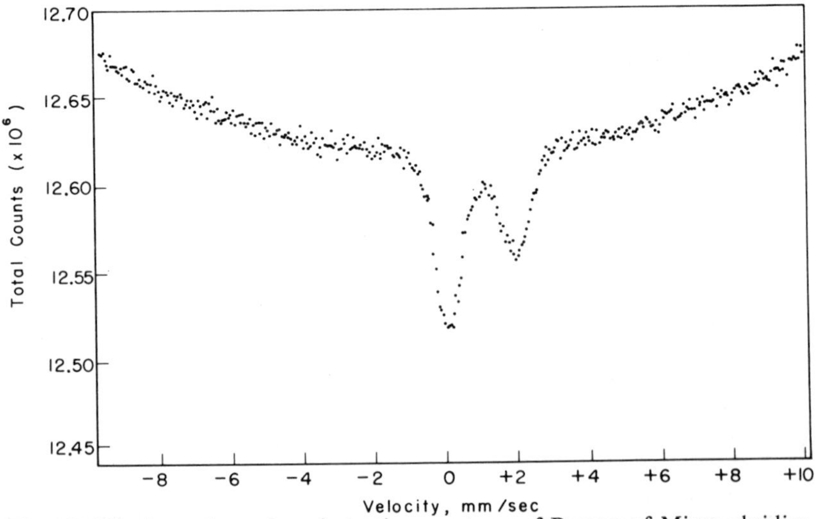

Fig. 10. Mössbauer hyperfine absorption spectrum of Bureau of Mines obsidian SLR 1B-10.

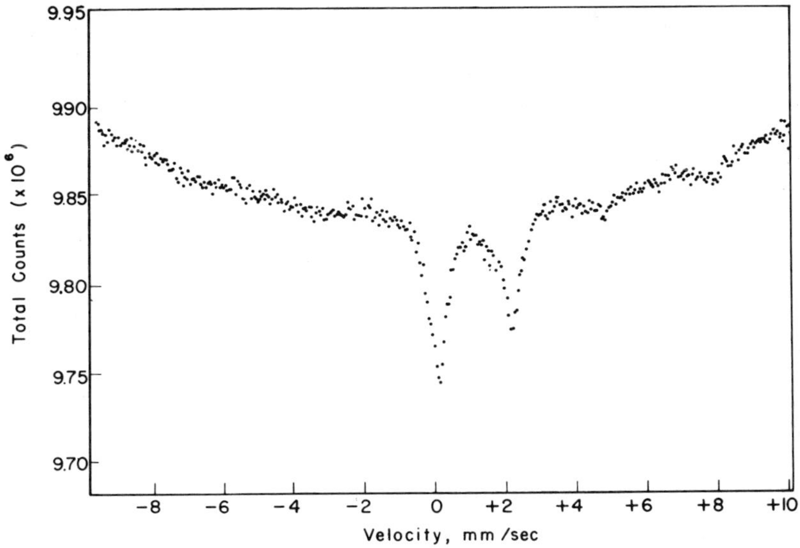

Fig. 11. Mössbauer hyperfine absorption spectrum of Bureau of Mines dacite SLR 1B-12.

Our earliest effort in this area was the examination of a suite of specimens from the Red Sea thermal brine regions.* These are high-iron sediments obtained from cores taken in several regions in the deeps of the Red Sea, in which sedimentary heavy-metal deposits appear to be actively in the process of forming at present. The particular samples which we investigated had already been studied using X-ray diffraction techniques [27], and in most cases studies of the magnetic properties of the samples had also been made [28]. In our experience Mössbauer spectrometry supplemented these techniques in a most useful fashion. We were able to verify the presence of major mineral phases, and in some cases detect minor mineral phases which had not been previously seen. Of special interest was the fact that in several cases we were able to verify the small grain size of certain phases which had been suggested by the magnetic measurements, and in some of these cases it is clear that one can, with the help of Mössbauer spectrometry, get better values for the typical crystallite size.

Figure 12 is a spectrum of one of the very first Red Sea sediment specimens that we looked at. This is from the Atlantis II deep, some

*We wish to thank D. W. Strangway, B. E. McMahon, and J. L. Bischoff for making available to us the samples and unpublished data.

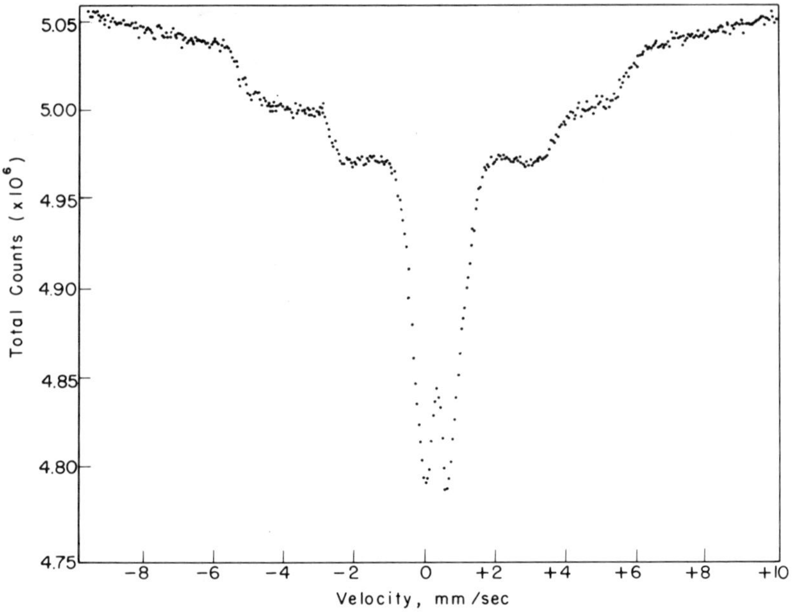

Fig. 12. Mössbauer hyperfine absorption spectrum of Red Sea core sample 127-455.

2000 m below the surface of the Red Sea, the depth in the core itself being 4.5 m. This is from the 2–62-μ fraction, these spherical aggregates being composed of smaller crystals. Goethite (α-FeOOH) identification was made on the basis of X-ray analysis. Some aspects of the thermal behavior of the sample supported this, but there were anomalies in both the thermal and magnetic behavior (e.g., the Curie-temperature effect at 120°C was not observed) which had led to the supposition that the material was so finely divided that it was behaving like paramagnetic material. The Mössbauer spectrum clears up the matter very nicely. There is obviously magnetically ordered material present, and the spectrum of bulk goethite is just as evidently absent (it would contribute a sharp six-line spectrum with the outermost peaks in the vicinity of ± 6 mm/sec) [12, 29]. In fact, apart from a contribution to the inner peaks from another mineral phase, this spectrum corresponds to what one would expect from finely crystallized superparamagnetic goethite, and comparison with measurements of the spectra of finely divided goethite which have appeared in the literature [11, 30, 31] suggest a distribution of crystallite sizes with typical dimensions of the order of 200 Å. X-ray studies on this sample had indicated the presence, in addition, of amorphous material. In all of the marine sediment specimens we have studied in which this X-ray amorphous material has been reported we have found this characteristic ferric doublet spectrum, which can be seen here as the main contribution in the paramagnetic region, although also present is a smaller doublet with slightly different parameters from the superparamagnetic goethite. The X-ray amorphous iron contribution has parameters identical, within experimental errors, to the parameters of the ferric oxide hydrate gels studied by van der Giessen [32], and the synthetic iron core polymers studied by Brady *et al.* [33]. We think that this is significant, and that the X-ray amorphous material which is such a significant component of these sediments is in fact probably a gel or polymer similar to these materials which have been prepared in the laboratory. Furthermore, we think we have evidence that such material is a precursor in the formation of goethite in natural marine environments, and we are planning further measurements to substantiate this.

Figure 13 is one of the spectra we have measured for samples from a suite of manganese nodules and associated sediments and their separated mineral fractions.* Manganese nodules are concretions which have formed in large quantities on the sea bed and in other marine and freshwater localities as well. They actually constitute one of the major types of rock encountered on the ocean floor. They commonly contain large amounts of both manganese and iron minerals, as well as other heavy metals. We have looked at several deep-sea specimens, as well as separated minerals from them. This is a specimen containing 10 Å manganite, which we think contributes to the inner peaks, and it appears that they may also contain a significant contribution from the colloidal or polymeric hydrated iron oxide form which I just referred to as a possible precursor in the formation of goethite. A paper which recently appeared in the literature on Mössbauer spectra from some manganese nodules also reported ferric peaks for the samples studied in about the same locations as we see them here [34]. This specimen is also of interest because there is also clear evidence in the Mössbauer spectrum of the presence of ferrous iron, which I believe is uncommon in manganese nodules. We estimated that the peak indicates the presence of about 0.25% by weight of ferrous iron in this manganese nodule.

*We wish to thank Dr. G. O. Arrhenius for making these specimens available to us.

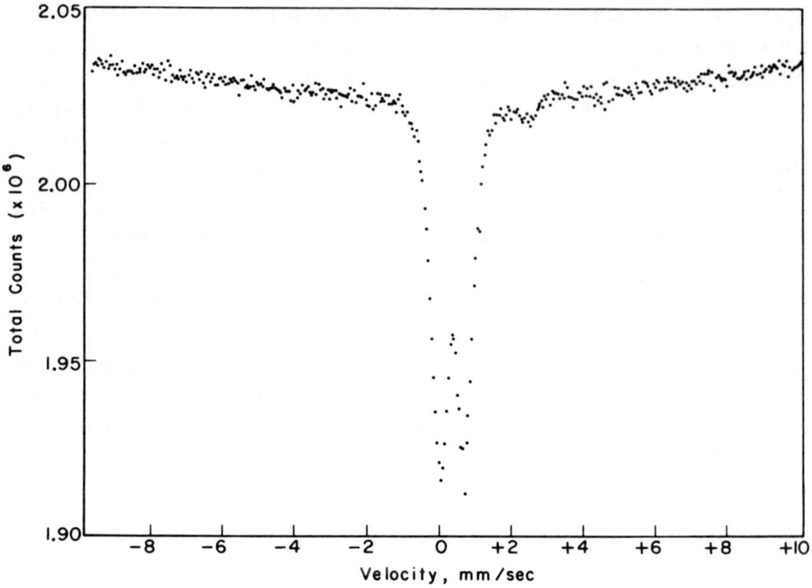

Fig. 13. Mössbauer hyperfine absorption spectrum of manganese nodule No. 17884.

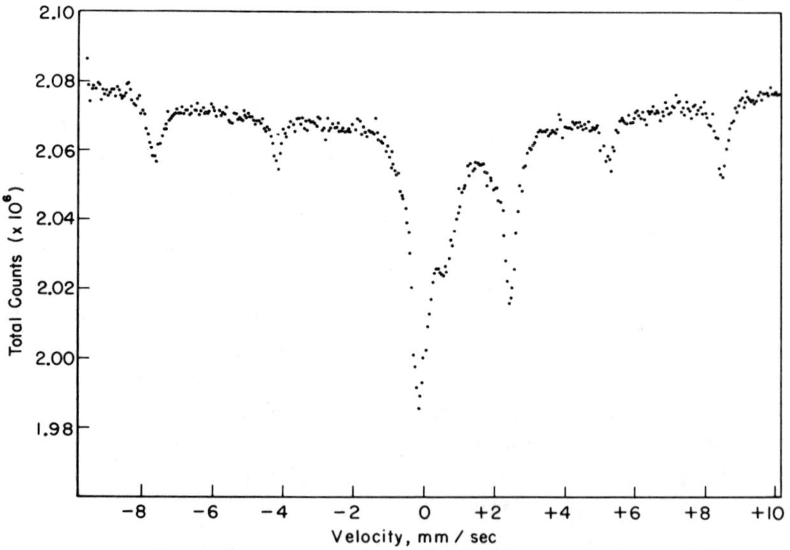

Fig. 14. Mössbauer hyperfine absorption spectrum of Atlantic red clay V16-213-276.

Figure 14 is another marine sediment, an Atlantic clay which had been investigated at Woods Hole Oceanographic Institution.* Our estimate of total iron of 5.3 % in this specimen was in reasonably good agreement with the results of chemical analysis. The red color of these clays had been surmised to be due to an iron oxide mineral coating on the clay particles, a coating which was not concentrated enough to be picked up by standard X-ray diffraction. As can be seen from the figure, Mössbauer spectrometry clearly indicates the presence of ferric oxide, and, in fact, the line positions are identical with those observed for the γ-phase of Fe_2O_3, which is the mineral maghemite. We understand that this result is quite an interesting one, and may provide a clue to the source area of the sediments, possibly the red beds of Nova Scotia. The positions of the inner lines of the paramagnetic spectrum suggested chlorite, which we found out later had shown up in the X-ray diffraction data, this being one of the detrital phases.

SUMMARY

In conclusion, I think it is clear that Mössbauer spectrometry can be introduced profitably as an instrumental technique for determinative mineralogy. It appears that it will be possible to obtain useful quantitative results without excessive difficulty. Our experience has been that this technique offers certain unique advantages and that the information obtained by its use can supplement that from other standard techniques in a highly satisfactory fashion.

REFERENCES

1. G. M. Bancroft, A. G. Maddock, and R. G. Burns, *Geochim. Cosmochim. Acta* **31**:2219 (1967).
2. S. S. Hafner, in: *Applications of Nuclear and Electronic Resonance Spectroscopy in Mineralogy*, American Geological Institute, Short Course Lecture Notes (American Geological Institute, Washington, D.C., 1968).
3. S. Ghose and S. S. Hafner, in: *Applications of Nuclear and Electronic Resonance Spectroscopy in Mineralogy*, American Geological Institute Short Course Lecture Notes (American Geological Institute, Washington, D.C., 1968).
4. A. S. Marfunin, R. M. Mineeva, A. R. Mkrtchyan, Ya. M. Niussik, and V. E. Fedorov, *Isv. Akad. Nauk SSSR, Ser. Geol.*, **10**:86 (1967).
5. S. Ghose and S. Hafner, *Z. Krist.* **125**:1 (1967).
6. G. M. Bancroft and R. G. Burns, *Earth Planetary Sci. Letters* **3**:125 (1967).
7. C. L. Herzenberg and D. L. Riley, unpublished research.
8. A. H. Muir, Jr., in: *Mössbauer Effect Methodology*, Vol. 4, I. J. Gruverman, ed. (Plenum Press, New York, 1968).
9. E. L. Sprenkel-Segel and S. S. Hanna, *Geochim. Cosmochim. Acta* **28**:1913 (1964).

*We wish to thank Dr. J. L. Bischoff for making Atlantic clay specimens and analyses available to us.

10. E. L. Sprenkel-Segel and G. J. Perlow, *Icarus* **8**:66 (1968).
11. J. G. Marzolf, J. T. Dehn, and J. F. Salmon, in: *The Mössbauer Effect and its Applications in Chemistry*, R. F. Gould, ed. (American Chemical Society, Washington D.C., 1967).
12. A. Lerman, M. Stiller, and E. Hermon, *Earth Planetary Sci. Letters* **3**:409 (1967).
13. B. W. Christ and P. M. Giles, in: *Mössbauer Effect Methodology*, Vol. 3. I. J. Gruverman, ed. (Plenum Press, New York, 1967).
14. P. A. Flinn, in: *Experimental Methods of Materials Research*, H. Herman, ed. (Interscience, New York, 1967.)
15. F. J. Flanagan, *Geochim. Cosmochim. Acta* **31**:289 (1967); **33**:81 (1969).
16. F. Chayes, *Geochim. Cosmochim. Acta* **31**:463 (1967).
17. J. Green, *Ann. N.Y. Acad. Sci.* **123**:1123 (1965).
18. D. E. Fogelson, "Simulated Lunar Rocks," US Bureau of Mines Report (1968).
19. G. Shirane, D. E. Cox, W. J. Takei, and S. L. Ruby, *J. Phys. Soc. Japan* **17**:1598 (1962).
20. S. K. Banerjee, W. O'Reilly, T. C. Gibb, and N. N. Greenwood, *J. Phys. Chem. Solids* **28**:1323 (1967).
21. S. K. Banerjee, W. O'Reilly, and C. E. Johnson, *J. Appl. Phys.* **38**:1289 (1967).
22. D. W. Collins, J. T. Dehn, and L. N. Mulay, in: *Mössbauer Effect Methodology* Vol. 3, I. J. Gruverman, ed. (Plenum Press, New York, 1967.)
23. A. A. Belyustin, Yu. M. Ostanevich, A. M. Pisarevskii, S. B. Tomilov, U. Bai-Shi, and L. Cher, *Soviet Phys.—Solid State* **7**:1163 (1965).
24. J. P. Gosselin, U. Shimony, L. Grodzins, and A. R. Cooper, *Phys. Chem. Glasses* **8**:56 (1967).
25. C. R. Kurkjian and D. N. E. Buchanan, *Phys. Chem. Glasses* **5**:63 (1964).
26. G. Tomandl, G. H. Frischat, and H. J. Oel, *Glastechn. Ber.* **8**:293 (1967).
27. J. L. Bischoff, in: *Hot Brines and Recent Heavy Metal Deposits in the Red Sea*, E. T. Degens and D. A. Ross, eds. (Springer-Verlag, New York, 1969).
28. D. W. Strangway, B. E. McMahon, and J. L. Bischoff, in: *Hot Brines and Recent Heavy Metal Deposits in the Red Sea*, E. T. Degens and David A. Ross, eds. (Springer-Verlag, New York, 1969).
29. J. B. Forsyth, I. G. Hedley, and C. E. Johnson, *J. Phys. C* (*Proc. Phys. Soc.*) **1**:179 (1968).
30. T. Shinjo, *J. Phys. Soc. Japan* **21**:917 (1966).
31. A. M. Van de Kraan and J. J. Van Loef, *Physics Letters* **20**:614 (1966).
32. A. A. van der Giessen, *J. Phys. Chem. Solids* **28**:343 (1967).
33. G. W. Brady, C. R. Kurkjian, E. F. X. Lyden, M. P. Robin, P. Saltman, T. Spiro, and A. Terzis, *Biochemistry* **7**:2185 (1968).
34. H. M. Gager, *Nature* **220**:1021 (1968).

Mössbauer Experiments with a He^3/He^4 Dilution Refrigerator*

G. M. Kalvius

Department of Technical Physics
Technical University of Helsinki
Otaniemi, Finland
and
Argonne National Laboratory
Argonne, Illinois

and

T. E. Katila and O. V. Lounasmaa

Department of Technical Physics
Technical University of Helsinki
Otaniemi, Finland

The large cooling capacity of He^3/He^4 *dilution refrigerators has made the temperature region below some hundredths of a degree available for Mössbauer resonance experiments. A spectrometer which allows the cooling of either the source or the absorber or both is described. Temperatures as low as* 0.03°K *were reached with this system. The application of very low temperatures is of interest for certain solid-state properties, such as low-lying magnetic transitions, paramagnetic relaxation, and localized magnetic moments, including the Kondo effect. It further allows Mössbauer measurements on oriented nuclei, which leads to an absolute thermometer, and gives information concerning the hyperfine interaction at the parent nuclei and the β-decay matrix elements. In addition to a general discussion of these applications, data which were obtained with the Mössbauer isotopes* Fe^{57}, Eu^{151}, *and* Au^{197} *will be presented.*

*Based on work performed in part under the auspices of the Finnish Atomic Energy Commission and the US Atomic Energy Commission.

INTRODUCTION

Almost all Mössbauer experiments at low temperatures published so far have been performed in cryostats in which liquid He^4 is the ultimate cooling agent. An early exception was the investigation of nuclear polarization of Co^{57} in iron down to temperatures of 0.35°K in a He^3 refrigerator by the Los Alamos group [1–4]. Recently Shinohara *et al.* [5] have used magnetic cooling to reach 0.025°K.

For many possible applications of the Mössbauer effect the temperature that can be obtained in a He^3 cryostat is not low enough. The adiabatic demagnetization technique suffers from a serious drawback, too: it is a "one-shot" method, which means the temperature cannot be kept constant during the relatively long counting times required in a Mössbauer experiment, especially in the presence of the large heat leak one encounters with cooled sources. Thus, the use of a Mössbauer demagnetization cryostat has been limited to absorber experiments with Fe^{57}, where a good spectrum can be obtained within an hour or so and the source is kept outside the cryostat at room temperature.

The recent advent of the He^3/He^4 dilution refrigerator [6–8] has brought a new and promising method for cooling and has been used successfully by us for Mössbauer experiments. A cryostat of this type is a continuously operating refrigerator capable of removing large amounts of heat even with modest pumping arrangements. In fact, once the necessary "know how" has been mastered, a dilution machine is simpler in construction and easier to operate than a demagnetization cryostat. The lowest temperature using a dilution refrigerator has been obtained by Wheatley *et al* [9], who reached 0.010°K in continuous operation and 0.0045°K in a "one-shot" process. Several groups have reported 0.025°K in continuous operation.

There are many reasons for pursuing Mössbauer experiments to very low temperatures. We should first mention studies of hyperfine interactions as a function of temperature in substances with magnetic transitions well below 1°K. Second, experiments which make use of nuclear polarization can only be conducted at temperatures at which the thermal energy kT is comparable with the hyperfine interaction energy. For most practical cases significant polarization occurs only at temperatures below 0.3°K. A typical application of this effect is the Mössbauer thermometer. A valuable feature of polarization experiments using the Mössbauer technique is that macroscopic orientation of the nuclei is not necessary. Finally, we should mention relaxation phenomena and localized moments in dilute alloys, in particular the Kondo effect, as possible candidates for Mössbauer studies at very low temperatures.

In this paper we shall first briefly explain the He^3/He^4 dilution refrigeration principle and then describe our cryostat. The rest of the paper is devoted to a discussion of experiments, most of which have been carried out with our apparatus.

THE DILUTION REFRIGERATION PRINCIPLE

To describe the principle of the dilution refrigerator we first review the properties of liquid mixtures of He^3 and He^4. Figure 1 shows the phase diagram in the (T, x)-plane, where x is the concentration of He^3. Above the phase-separation curve the liquid is superfluid on the left side of the λ-curve and normal fluid on the right side of the λ-curve. Below the phase-separation curve the liquid spontaneously divides itself into two components, one of the phases being rich in He^3 and the other rich in He^4. Because of its lower density, the He^3-rich phase floats on top of the He^4-rich phase. The (T, x)-relationship in the upper (concentrated He^3) phase is given by the right branch of the phase-separation curve; similarly, the left branch of this curve represents the lower (dilute He^3) phase. Since the final temperature in

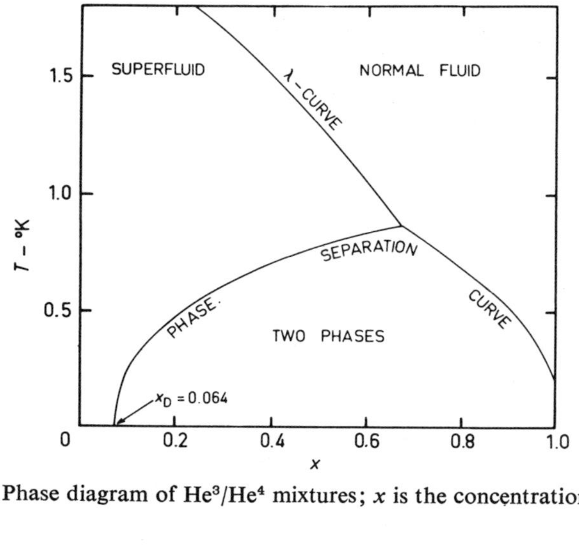

Fig. 1. Phase diagram of He^3/He^4 mixtures; x is the concentration of He^3.

a dilution refrigerator usually is below 0.1°K, we observe from Fig. 1 that in the concentrated phase $x_c \approx 1$ and that in the dilute phase x_D varies slightly with temperature, from 0.064 at 0°K to 0.070 at 0.1°K. Basically, the 6.4% solubility of He^3 into He^4 at $T = 0$ is caused by the fact that a single He^3 atom is more strongly bound into pure liquid He^4 than into pure He^3.

The principal parts of a dilution refrigerator are shown schematically in Fig. 2. The phase separation occurs in the mixing chamber, and cooling is produced there by moving He^3 atoms from the upper, almost pure He^3 phase across the boundary into the lower, dilute phase. This process is, in many respects, analogous to ordinary evaporation. The upper phase corresponds to the liquid and the lower phase to the vapor. The reason for the validity of this comparison is the fact that He^4, because of its zero nuclear spin and superfluid properties, is thermally and hydrodynamically inert at temperatures below 0.5°K and acts just as a "supporting" medium for He^3 in the dilute phase.

The continuous dissolving of He^3 atoms from the concentrated to the dilute phase is caused by circulating He^3 in the system by means

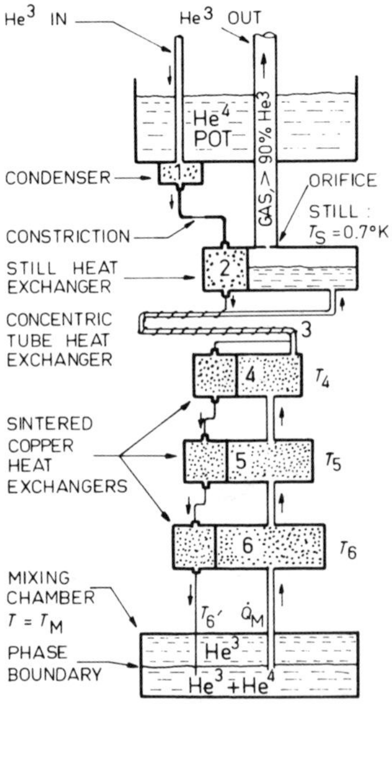

Fig. 2. Schematic drawing of the principal parts of our dilution refrigerator.

of a pump at room temperature. Incoming gas (Fig. 2) is first precooled in a condenser attached to the pumped He^4 bath at 1.2°K. The pressure of He^3 is kept sufficiently high for condensation to occur by means of a flow-limiting construction. The liquid then enters the still heat exchanger at 0.6–0.8°K. It continues through the main heat exchangers and finally reaches the mixing chamber. After crossing the phase boundary the He^3 atoms proceed again through the main heat exchangers, along an unbróken column of the dilute superfluid phase, to the still. Vapor is removed from the still by pumping. Over 90% of the outcoming gas is He^3, because He^4 atoms are, even at the still temperature, relatively inert. The small amount of He^4 in the circulating gas can be ignored in a simplified discussion of the dilution refrigerator. External heating of the still is required to keep its temperature high enough to achieve a sufficient gas circulation rate.

We next assume that the dilution refrigerator has reached a steady state, the temperatures of the still, the heat exchangers, and the mixing chambers (Fig. 2) being constant at T_s, T_4, T_5, T_6, and T_M, respectively. Figure 3 shows $H_{3C}(T)$, the enthalpy per mole of pure He^3 as a function of temperature, and $H_{3D}(T, x)$, the enthalpy per mole He^3 of the dilute phase as a function of temperature and the appropriate He^3 concentration. The variation of x with T is given by the requirement that the chemical potential is constant along the dilute side. If all heat loads and effects due to viscosity are neglected, the enthalpy on the concentrated side will decrease in a given heat exchanger with the same amount as the enthalpy on the dilute side increases. The function of the heat exchangers can thus be represented, starting from $H_{3C}(T_S)$, as illustrated in Fig. 3. Finally, in the mixing chamber we have a constant-enthalpy process which reduces the temperature roughly by a factor of 2.8. The limiting temperature of an ideal dilution refrigerator with three isothermal heat exchangers is thus ~0.010°K. By increasing the number of heat exchangers lower temperatures can be reached.

In an actual refrigerator external heating must be taken into account. The heat-balance equation reads at the mixing chamber

$$\dot{Q}_M = \dot{N}_3 \, \Delta H = \dot{N}_3[H_{3D}(T_M) - H_{3C}(T_6')] \tag{1}$$

Here $\dot{Q}_M = dQ_M/dt$ is the external heating rate, $\dot{N}_3$ is the molar circulation rate of He^3, and T_6' is the temperature at which He^3 leaves the last heat exchanger. The maximum cooling power is achieved if

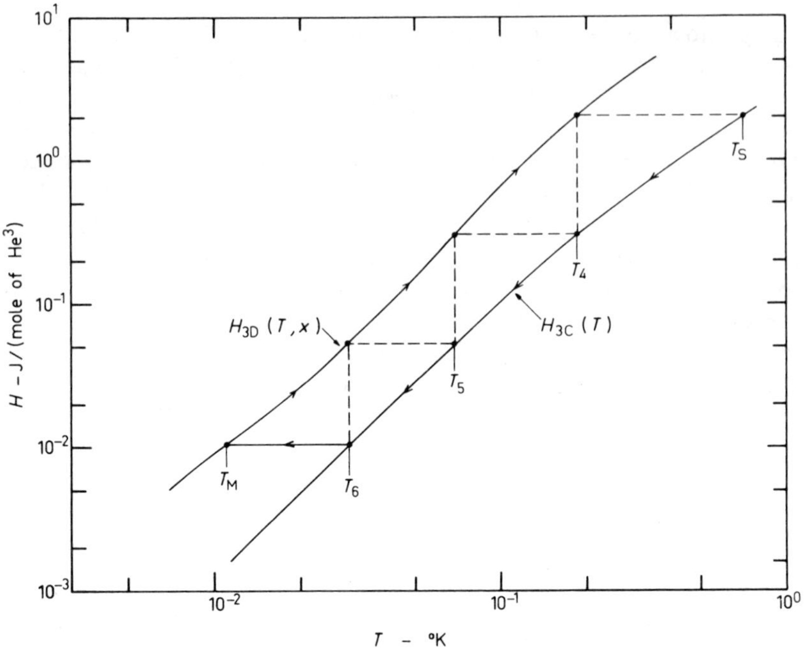

Fig. 3. Plots of $H_{3C}(T)$ and $H_{3D}(T, x)$, the enthalpies of the concentrated and dilute sides, respectively, as a function of temperature.

$T_M = T_6'$. Experimental measurements of H_{3C} and H_{3D} show that in this case, below 0.04°K, Eq. (1) has the form

$$\dot{Q}_M = 83\dot{N}_3 T_M{}^2 \quad \text{W}\cdot\text{sec/(mole of He}^3)\cdot{}^\circ\text{K}^2 \qquad (2)$$

We now clearly see the difference between the dilution process and an ordinary evaporation-type refrigerator. In the latter case a limiting temperature is reached because the vapor pressure decreases as $p_v \propto \exp(-\alpha/T)$, where α is a positive constant. Thus, fewer and fewer molecules can be removed by the pump, and eventually a stable situation occurs at which cooling due to evaporation just balances the external heat leak. In a dilution refrigerator, when the temperature of the mixing chamber is lowered during the operation of the cryostat, we do not "run out" of He^3 molecules in the dilute or "vapor" phase; there will be 6.4% of He^3 even at $T = 0$. That means $\dot{N}_3$ does not change with T at low temperatures, and thus, according to Eq. (2), the cooling capacity of a dilution refrigerator decreases as T^2, not as

$\exp(-\alpha/T)$. This is a very important advantage over an ordinary evaporation-type cryostat.

By assuming $\dot{Q}_M = 3$ ergs/sec and $\dot{N}_3 = 3 \times 10^{-5}$ mole/sec, we obtain from Eq. (2) $T_M = 0.011°K$. This calculation shows the capabilities of a dilution refrigerator. Even with a modest gas circulation rate, easily handled by a small pumping system, and with a considerable heat leak present, temperatures below 0.02°K can be reached.

For more details about matters discussed in this section we refer to a paper by Lounasmaa [10] and to other publications cited therein.

THE CRYOSTAT

The most important construction details of our dilution cryostat for Mössbauer effect research are shown in Fig. 4. The low-temperature parts can be divided into two separate units: the Dewar and the insert. The Dewar is a modification of a cryostat described earlier by Kalvius [11]. It is now commercially available.* Its central feature is a 3-in. o.d. stainless steel tube which runs vertically through the main He^4 reservoir. The lower half of this tube is made of copper, with a conical upper end. The insert is firmly supported by and thermally anchored to this cone. The γ-radiation emerges through beryllium or Mylar windows and an aluminum-foil radiation shield at the bottom of the Dewar.

The principal parts of the insert are the pumping head, the vibrator,† the spring-operated thermal anchor at 80°K, the conical thermal anchor at ~7°K, the He^4 pot, which can be pumped to 1.2°K, the vibrating rod (made of $\frac{3}{8}$-in. o.d., 0.01-in.-thick stainless steel tube, guided by two beryllium bronze springs), and finally the various parts of the dilution refrigerator. During experiments the operating space is evacuated. A special feature of our design is that the heat leak from above is entirely absorbed by the thermal anchors and by the He^4 pot; there is no main liquid bath through which the various tubes pass. This is to avoid separate vacuum spaces into which the vibrating rod would have to enter. The insert can be pulled out of the Dewar and replaced in a few minutes. The source and the absorber are thus easily accessible.

The main parts of the dilution refrigerator itself (Figs. 2, 4, and 5) are the still, the mixing chamber, and the various heat exchangers.

*Model 10DT Mössbauer Cryostat by Janis Research Corporation.
†Model K3 of Austin Science Associates, Austin, Texas.

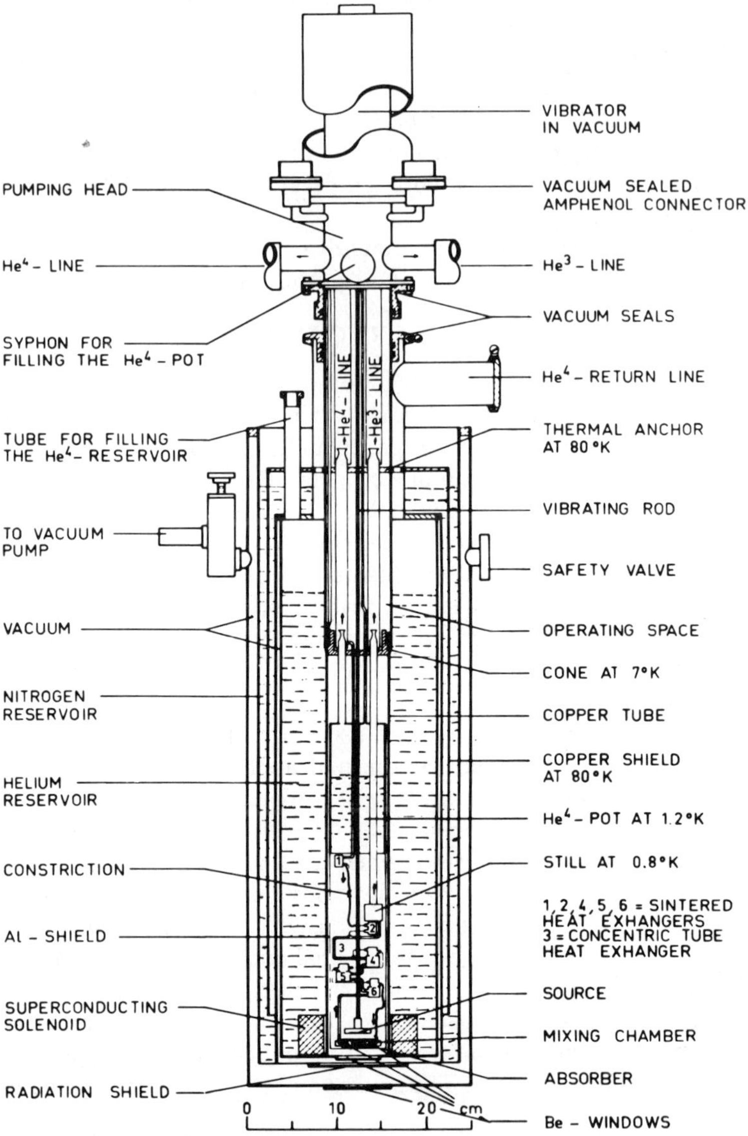

Fig. 4. Schematic illustration of our cryostat.

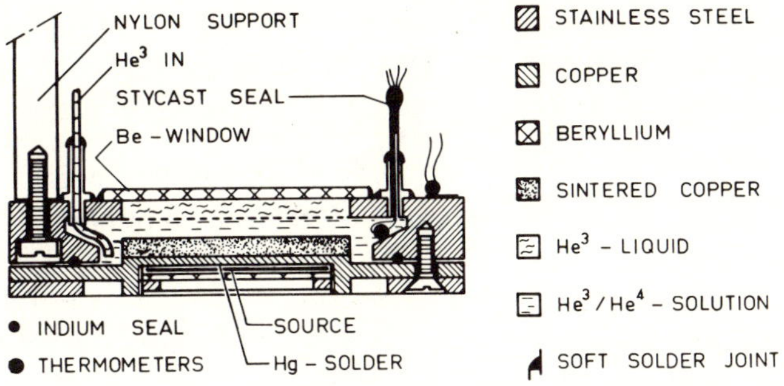

Fig. 5. One typical construction of the mixing chamber.

The still is a 2 cm^3 brass chamber with a copper bottom. The mixing chamber used in source experiments is shown in Fig. 5. When absorbers are to be cooled they are put into the mixing chamber and the lower copper plate is replaced by a beryllium window. The chamber volume is about 4 cm^3.

We have used two main constructions of heat exchangers, the sintered copper type and the concentric-tube type. All heat exchangers except No. 3 (Fig. 4) are of the sintered type. These were made by sintering copper powder (particle size $<44\ \mu$) for 40 min at 800°C in a hydrogen atmosphere. Firm bonding to the copper walls was achieved by pressing the powder with a graphite rod during the sintering process. The concentric-tube heat exchanger No. 3 is located immediately below the still. It is constructed by placing a CuNi tube 40 cm long and 0.43/0.27 mm in diameter inside another tube 1.0/0.8 mm in diameter, of the same length and material, and by winding the unit into a horizontal helix.

When the absorber is to be cooled to the lowest temperatures the source is mounted at the end of the vibrating rod (Fig. 4). It will normally have a temperature around 1.5°K. If it is desired to cool the source the absorber is attached to the vibrating rod by means of a fork which allows it to be moved below the mixing chamber.

The maximum He^3 circulation rate in our dilution cryostat is about 3×10^{-5} mole/sec. The machine can be cooled to slightly below 0.03°K, and operating times up to 24 hr are possible with one filling of the He^4 pot. Calibrated germanium and carbon resistance thermometers are used to measure the mixing chamber temperature. A feedback circuit driven by the thermometer in the mixing chamber supplies power into a heater for stabilizing the temperature. The circuits used for temperature readout and regulation are similar to the ones described by Steyert and Daybell [12].

At low temperatures a serious problem is radioactive heating. If the source is to be cooled inside the mixing chamber, the radioactive self-heating and the Kapitza thermal boundary resistance will prevent it from reaching the temperature of the surrounding liquid. This difference in temperature is roughly proportional to $1/T^3$. Using measured values of the Kapitza resistance one can show that it is impossible to cool a Co^{57} source of a few millicuries with a practical surface-to-volume ratio much below 0.1°K. More effective cooling can be obtained by using the construction shown in Fig. 5. For a 10-mCi Co^{57} source the total heating power is 80 ergs/sec, a load which, if absorbed by the mixing chamber, is sufficient for preventing the refrigerator from cooling below 0.07°K.

In absorber experiments radioactive heating is generally very small, and calculations show that one should be able to reduce the temperature at least an order of magnitude further. It is necessary, however, to shield the source in such a way that most of the heat is absorbed, for instance, by the still, and to make sure that radiation does not strike the mixing chamber outside the area of the absorber. Sources which emit β- or soft X-rays must be shielded from the absorber by an aluminum foil. The use of enriched absorbers helps further to reduce the heating effects, since the electronic absorption of gamma rays is reduced.

For more details about the construction and operation of our dilution refrigerator we refer to a paper by Ehnholm *et al.* [13].

We now turn our attention to several different types of experiments carried out with our cryostat.

THE MÖSSBAUER EFFECT THERMOMETER

In the presence of a hyperfine field a nuclear state separates into a series of sublevels. As an example, the ground state of Au^{197}, which has a spin $I = \frac{3}{2}$, will split in an internal magnetic field H_i into four hyperfine levels, characterized by the magnetic quantum number m_I (Fig. 6). The magnetic interaction energy is $-\mu_I H_i m_I/I$. Here $\mu_I = g_I \mu_N I$ is the magnetic moment of the nuclear state with spin I, and μ_N is the nuclear magneton. Thus, we have for the separation of two successive levels ($\Delta m_I = 1$)

$$\mathscr{E} = -\mu_I H_i/I = -g_I \mu_N H_i \tag{3}$$

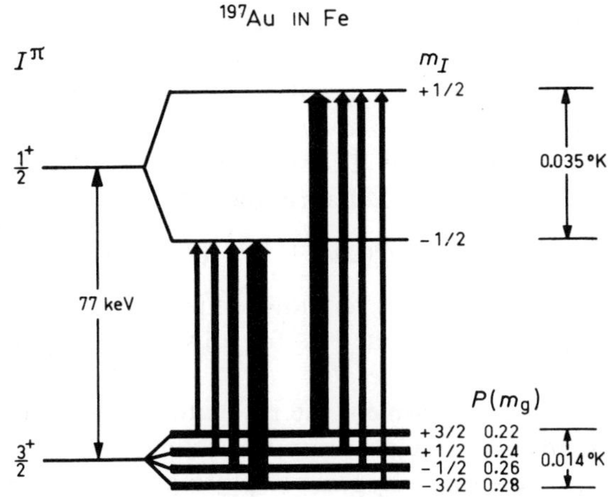

Fig. 6. Hyperfine transitions of Au^{197} in a magnetic field. The widths of the arrows indicate their relative line intensities at 0.06°K.

The total magnetic splitting of the nuclear state I is $2\mu_I H_i$. The population of each level is given by the Boltzmann factor,

$$P(m_i) = \exp[-m_I \mathscr{E}/kT] / \sum_{m_I} \exp[-m_I \mathscr{E}/kT]$$

$$= \exp[m_I g_I \mu_N H_i/kT] / \sum_{m_I} \exp[m_I g_I \mu_N H_i/kT] \tag{4}$$

At low temperatures, where the thermal energy becomes comparable with the separation of the extreme hyperfine levels, i.e., when

$$kT \approx 2\mu_I H_i \tag{5}$$

the individual sublevels will no longer be equally populated. In our example the gold nuclei are placed in a matrix of metallic iron. A hyperfine field $H_i = -1.28$ MOe will be present at the site of Au^{197}, causing a total ground-state splitting equal to 0.014°K (Fig. 6). If the sample temperature is of this order of magnitude, the $m_I = +3/2$ state will be less populated than the $m_I = -3/2$ state; nuclear polarization is thus observed. It should be pointed out that polarization is not limited to magnetic-dipole interactions as discussed above; any type of nuclear hyperfine splitting will be sufficient. For instance, the pure quadrupole interaction energy is $e^2qQ[3m_I^2 - I(I+1)]/4I(2I-1)$ for a rotationally symmetrical electric field gradient q interacting with the nuclear quadrupole moment Q. Since the levels in this case are characterized by m_I^2 rather than by m_I, one speaks more correctly of nuclear alignment. The ground state of our example, Au^{197}, would split into two levels characterized by the quantum numbers $m_I = \pm 1/2$ and $m_I = \pm 3/2$, respectively. The separation of two successive levels m_I and $m_I + 1$ in the case of quadrupole splitting is

$$\mathscr{U} = 3e^2qQ(2m_I + 1)/4I(2I - 1) \tag{6}$$

Usually, $\mathscr{U}$ is quite small, and very low temperatures are thus required with pure quadrupole interaction for observing a change in the population of the hyperfine levels.

In a Mössbauer experiment the resonantly absorbed gamma ray will excite an absorber nucleus from its ground state I_g to its excited state I_e. In the presence of hyperfine splitting this process will occur between the sublevels m_g and m_e. The relative absorption A due to such a transition, as observed in a Mössbauer spectrum, is proportional to

$$A(m_g \to m_e) \propto P(m_g)C(m_g, m_e) \tag{7}$$

where $C(m_g, m_e)$ is the square of the appropriate normalized Clebsch–Gordan coefficient and $P(m_g)$ is defined by Eq. (4). In the presence of nuclear polarization the intensities of the lines in the spectrum will be affected, *via* $P(m_g)$, by the Boltzmann factors. Consequently, the absorption A can directly be related to the absorber temperature. In the case of a pure magnetic interaction the hyperfine levels are fully characterized by m_I. Since $C(m_g, m_e) = C(-m_g, -m_e)$, we obtain, by combining relations (3), (4), and (7), for the intensity ratio ρ of the two lines in a Mössbauer spectrum which originate from the extreme ground-state levels

$$\rho = \frac{A(+m_g \to +m_e)}{A(-m_g \to -m_e)} = \frac{P(+m_g)}{P(-m_g)} = \exp\left(\frac{-2m_g\mathscr{E}}{kT}\right) \tag{8}$$

(In the case of our example in Fig. 6 we will choose the lines: $+3/2 \to +1/2$ and $-3/2 \to -1/2$). It should be noted that $\mathscr{E}$ can be obtained at the same time from the separation of the resonance lines. We thus have a thermometer which supplies its own calibration; as soon as $\mu_I H_i/I$ has been determined from the line separation the intensity ratio ρ gives the absolute temperature.

Figure 7 shows a typical example of nuclear polarization effects in Au^{197}. At 0.043°K a reduction in the line intensities at higher

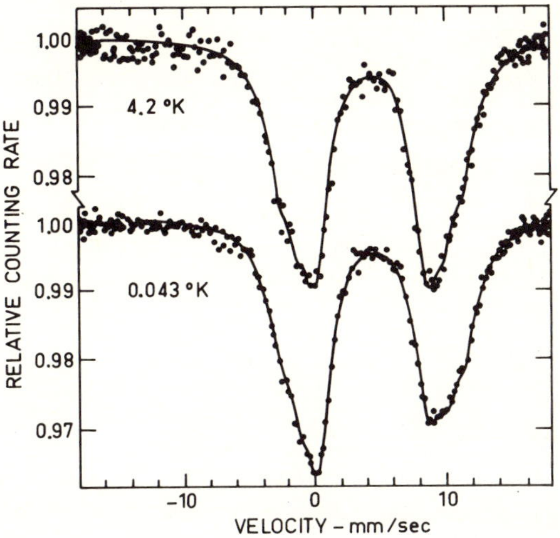

Fig. 7. Mössbauer spectra of Au^{197} for an absorber of 1 at. % gold in iron at 4.2° and 0.043°K. The source is Pt^{197} in platinum at 1.5°K. The polarization effects are clearly visible in the lower spectrum.

energies (positive velocities) is very evident. The hyperfine field in the absorber was produced by alloying 1% gold into iron metal [14]; a single-line source (Pt^{197} in platinum metal) was employed.

For Fe^{57} the magnetic dipole moment of the ground state is unusually small: $\mu_{1/2} = +0.0902$ nm. The largest hyperfine fields obtainable in iron compounds are around 600 kOe. This leads to a ground-state separation of ~0.004°K. Very low temperatures are thus required before polarization effects become evident in an Fe^{57} absorber experiment. Figure 8 shows the spectrum of Fe_2O_3 observed at 0.043°K with about 8% polarization [15]. It is quite possible that the Fe^{57} Mössbauer effect absorber thermometer will prove useful for calibration purposes and for investigations of the cerium magnesium nitrate (CMN) magnetic temperature scale in the vicinity of 0.01°K and below.

Several other transitions could, in principle, be used as a Mössbauer effect absorber thermometer. Some of them are listed in Table I. In Fig. 9 we show the spectrum [16] of the 21.6-keV transition of Eu^{151} in EuS at 0.054°K. The polarization is very considerable, owing to the large total splitting of the ground state (Table I).

Although the Mössbauer absorber thermometer is theoretically very simple, its practical realization is connected with a number of

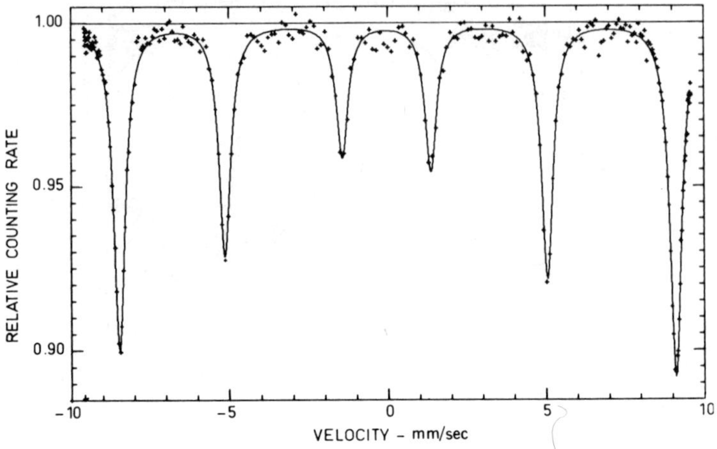

Fig. 8. Mössbauer resonance of Fe^{57} in Fe_2O_3 at 0.043°K; the source is Co^{57} in copper at 1.5°K. The polarization is ~8%.

Table I. Properties of Isotopes Usable as Mössbauer Absorber Thermometers

Isotope	Transition energy, keV	Ground-state magnetic moment, nm	Absorber material	Hyperfine field, kOe	Splitting factor $2\mu_g H_i/k$, °K	Spectrum
Np^{237}	59.6	+2.70	$NpAl_2$	3150	0.621	Resolved
Dy^{161}	25.7	−0.472	Dy metal	7100	0.240	Resolved
Yb^{171}	66.7	+0.493	$YbCl_3 \cdot 6H_2O$	2980	0.107	Resolved
Sb^{121}	37.2	+3.359	MnSb	353	0.087	Complex
Eu^{151}	21.6	+3.463	EuS	332	0.084	Complex
Ir^{193}	73.0	+0.159	1% Ir in Fe	1495	0.017	Partly resolved
Sn^{119}	23.9	−1.046	Sn ferrite	210	0.016	Resolved
Au^{197}	77.3	+0.146	1% Au in Fe	1280	0.013	Partly resolved
Fe^{57}	14.4	+0.090	FeF_3	620	0.004	Resolved

technical and other difficulties which limit its usefulness severely. For obtaining an accurate temperature reading it is necessary to measure line intensities (or preferably the areas under lines) with a small statistical error. Counts must thus be collected for a relatively long time, perhaps several hours. Rather sophisticated equipment, when compared with other types of thermometers, is required.

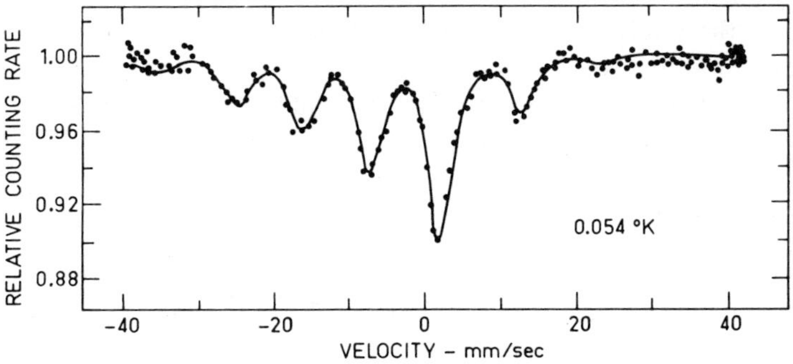

Fig. 9. Polarized spectrum of Eu^{151} in EuS at 0.054°K. The source is Sm_2O_3 at 1.5°K.

A more fundamental drawback is that the transmitted intensity at resonance is a function of t, the effective absorber thickness [17]. In Fig. 10 the apparent temperature, deduced from the ratio of either the peak intensities or the area of the outermost hyperfine lines for the Fe^{57} resonance in Fe_2O_3 at 0.0084°K, is plotted against the effective absorber thickness. Since the area under a line saturates somewhat more slowly than the peak intensity [18], it is advantageous to employ

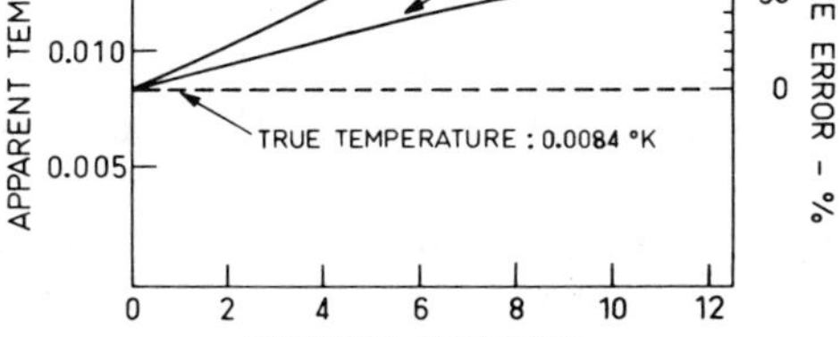

Fig. 10. Calculated effect of the absorber thickness on an absolute Mössbauer thermometer. The example is Fe^{57} in Fe_2O_3 at 0.0084°K. The absorber thickness is defined according to ref. [17]. The apparent temperature has been obtained by either taking the amplitude ratio or the ratio of the areas of the two outermost hyperfine lines.

areas in the analysis, as Fig. 10 clearly demonstrates. This method, however, requires that at least a partial velocity spectrum has been measured, whereas for obtaining the amplitude ratio it would, in principle, be sufficient to count only at two velocities which correspond to the peaks of the outermost lines. Even if the area method is used, a practical absorber ($t \approx 2$) will still cause an error of 15% in temperature if the finite absorber thickness is not taken into account. This correction becomes especially serious if the hyperfine lines are not well separated as in the case of Eu^{151}. The full transmission integral [17], instead of a sum of Lorentzians, must then be employed in the analysis of the spectrum. This involves a large computer and is expensive in computer time, and, in addition, requires a knowledge of t. This quantity is usually not well known, since the absolute recoil-free fraction in an absorber is hard to determine. Very thin absorbers are impractical because the statistical accuracy suffers, and good uniformity is hard to achieve. For precision measurements it is necessary to take data with several absorbers of different thicknesses and to extrapolate the results to $t \to 0$ by using the formulas of refs. [17, 18], or to calibrate the thermometer at a known temperature.

The thermometers described so far can be classified as "absorber thermometers," as we have used nuclear polarization effects which occur in the resonance absorber. In this case the source is a single-line emitter and may be at any temperature convenient for the experiment. It is, of course, also possible to utilize the nuclear polarization in a split source at very low temperatures. In this arrangement the absorber is unsplit and just serves as an analyzer of the spectrum. This setup, which we call a "source thermometer," avoids the errors caused by the finite absorber thickness, but, as we shall see, is quite complex, and, in particular, the highly desirable features of self-calibration is lost.

For a discussion of the source thermometer let us return to our example of Au^{197}. The parent nucleus is Pt^{197} and its simplified decay scheme is shown in Fig. 11. In the presence of a hyperfine field H_i the $I = 1/2$ ground state of Pt^{197} will split into two sublevels. The β-transitions will lead to the two hyperfine levels of the 77-keV first excited state of Au^{197} ($I = 1/2$). The gold nucleus subsequently reaches its ground state by gamma-ray emission. The hyperfine spectrum is then analyzed in a Mössbauer experiment.

The intensity of a transition from an excited level to the ground state is given, according to Eq. (7), by

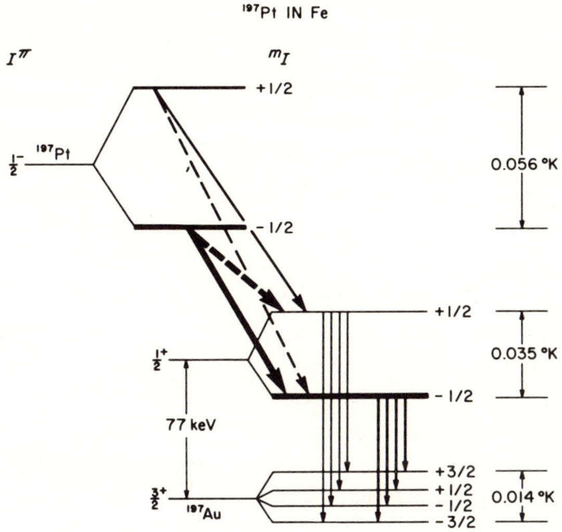

Fig. 11. Simplified decay scheme of Pt^{197} (a 10% β branch to a higher excited state has been omitted) in the presence of a hyperfine field at very low temperatures. The broken arrows represent β transitions with vector matrix elements. The width of the hyperfine levels indicates their population.

$$A(m_e \rightarrow m_g) \propto P(m_e)C(m_e, m_g) \tag{9}$$

where

$$P(m_e) = \sum_{m_p} P(m_p)W(m_p, m_e) \tag{10}$$

The last equation connects the population of the sublevel m_e in the daughter with the population of the sublevel m_p in the parent *via* the β-transition probability $W(m_p, m_e)$. In turn, $P(m_p)$ depends on the temperature through the relation [cf. Eq. (4)]:

$$P(m_p) = \exp[-m_p\mathscr{E}_p/kT]/\sum_{m_p} \exp[-m_p\mathscr{E}_p/kT] \tag{11}$$

Thus the level separation $\mathscr{E}_p = -\mu_p(H_i)_p/I_p$, which determines the polarization, depends on the dipole moment μ_p, on the ground state spin I_p of the parent nucleus, and on $(H_i)_p$, the hyperfine field to which the parent nuclei (Pt^{197} in our example) are subjected by their surroundings. However, $\mathscr{E}_p$ cannot be determined simultaneously in the Mössbauer experiment, since the separation of the observed resonance lines is obviously given by the hyperfine interaction in the daughter nucleus (Au^{197} in our example). Thus the self-calibration feature is lost and $\mathscr{E}_p$ has to be determined in a separate experiment. This may be done by measuring the polarization of the source at one known temperature.

Another difficulty arises because the β-transition probability $W(m_p, m_e)$ must be known. This problem will be discussed in more detail below. Let us state here that *a priori* the value of $W(m_p, m_e)$ is known only for allowed β-decays ($\Delta I = 0, \pm 1$; no parity change). In the β-decay of Pt^{197} a parity change is involved (Fig. 11), and thus this resonance does not well qualify as a source thermometer. Finally, in Eq. (10) it was tacitly assumed that the nuclear spins are not flipped during the decay process. Whether this assumption is valid will depend on the host lattice in which the source nuclei are placed. In metals, due to the presence of conduction electrons, the nuclear spin-relaxation times will be shorter than in insulators. However, at very low temperatures the relaxation times are generally longer than the lifetimes of the Mössbauer levels. Thus, depolarization effects will not be present. To obtain long spin-relaxation times, an insulator would be even more preferable; however, other complications will arise. For instance multiple Auger electrons emitted after a K-capture will produce a variety of electronic charge states. These may decay with times comparable to the lifetime of the Mössbauer level in insulators, and thus produce multiple resonance spectra. In metals these decay times are considerably shorter (10^{-14} sec). These "after-effects" in insulators make it impossible to use the Co^{57} thermometer to determine their temperature. This is unfortunate, since, due to their poor thermal conductivity, it is usually difficult, if not impossible, to determine the temperature of an insulator at very low temperatures by other methods.

It is fortunate that the 14.4-keV gamma ray of Fe^{57} is preceded by an allowed β-decay from Co^{57}. In addition, the large dipole moment of the ground state of Co^{57} ($\mu_p = +4.58$ nm) causes a total splitting of 0.102°K for cobalt in metallic iron. Such a source has been used earlier by the Los Alamos group [1–4] to demonstrate the feasibility of a Mössbauer thermometer. These authors discuss in detail the features of the Co^{57} thermometer, and some additional comments may also be found in ref. [12]. We have extended those measurements to lower temperatures [15]. A typical spectrum is shown in Fig. 12; the source was cooled to 0.13°K. The asymmetry is evident, and is particularly clear in the outermost pair of lines.

The usefulness of a Mössbauer-effect-source thermometer is limited principally for two reasons. First, the possibility of relaxation

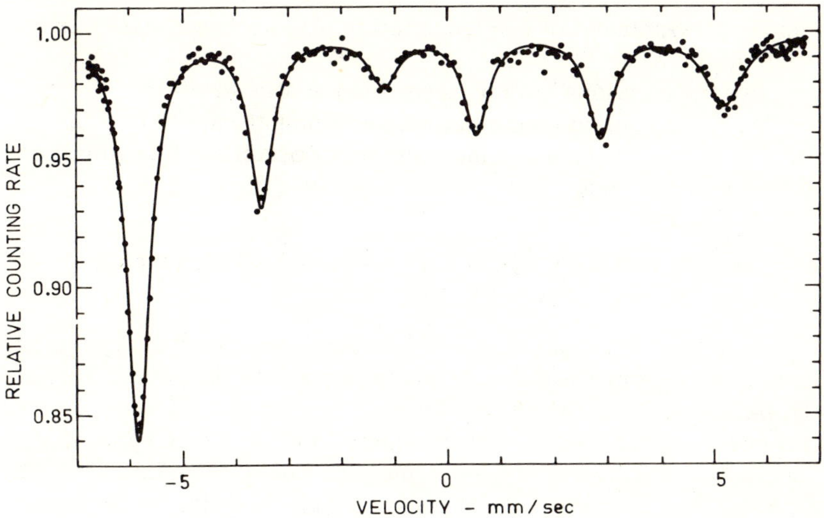

Fig. 12. The Co^{57}-source thermometer. A 3-mCi quantity of Co^{57} was diffused into an iron matrix and cooled to 0.04°K using the mixing chamber configuration shown in Fig. 5.

effects must always be kept in mind, as they could alter the intensity ratios completely. Second, radioactive self-heating, as discussed in the previous section, can prevent the source from reaching equilibrium with its surroundings at low temperatures. In the case of Co^{57} in iron, placed inside the mixing chamber, we found that the source did not cool below 0.12°K. It should be mentioned here that, in general, sources which decay by electron capture are more favorable than β-emitters. In the former case most of the decay energy is carried away by neutrinos and not dissipated inside the cryostat. In any case a candidate for a Mössbauer source thermometer should have a high ratio of resonance gamma rays to β-particles. This requires a large feeding ratio and a small conversion coefficient for the Mössbauer level.

We summarize the Mössbauer thermometer as follows: The absorber thermometer supplies its own calibration, but its accuracy is limited by thickness effects, which must be determined separately. The source thermometer requires a calibration at one temperature, suffers from the possibility of relaxation phenomena, and is susceptible to equilibrium difficulties due to radioactive self-heating. Both types of thermometer require rather sophisticated experimental equipment, and the measurement of one temperature may involve counting times up to an hour.

For these reasons it is not likely that the Mössbauer thermometer will become a common method for determining the absolute temperature. As mentioned before, it may play an important role for calibration purposes in the region between 0.001 and 0.02°K.

MEASUREMENTS OF HYPERFINE INTERACTION AT SOURCE NUCLEI

Equations (9) and (10) show how the intensity of the hyperfine lines emitted from a Mössbauer source depend on the polarization of the parent nucleus. In the preceding section we have discussed how this effect can be used as a thermometer. On the other hand, if the temperature is known, we may employ the polarization to determine $\mathscr{E}_p = \mu_p(H_i)_p/I_p$ and thus the hyperfine field $(H_i)_p$ acting on the parent nucleus. Because the β-transition probabilities $W(m_p, m_e)$ must be known, absolute measurements are again restricted to cases where the Mössbauer level is populated by an allowed β-decay. This restriction could be lifted if very low temperatures would be reached (as discussed at the end of the next section). The determination of the ratio of internal fields for two different source matrices is always possible. Furthermore, we must assume that the polarization is not affected by fast relaxation (see preceding section).

The most important case is again Fe^{57} with Co^{57} as the parent. Dash *et al.* [1, 2] have used the polarization technique for determining the hyperfine field at dilute cobalt in iron as $H_i(\text{Co}/\textbf{Fe}) = -(300 \pm 20)$ kOe. Recently Reivari [19] has studied the hyperfine interaction of dilute cobalt in palladium. His spectrum is shown in Fig. 13. A comparison with the ultra low temperature spectrum of Co^{57} in iron (Fig. 12) shows immediately that the sign H_i in palladium must be opposite to that in iron. The internal field was determined to be $H_i(\text{Co}/\textbf{Pd}) = +(280 \pm 20)$ kOe.

MIXING RATIO OF β-DECAY MATRIX ELEMENTS

In the preceding sections it was stated that a calculation of the β-transition probability $W(m_i, m_f)$ between two hyperfine levels can only be calculated for allowed transitions. They obey the selection rule $|I_i - I_f| = 0,1$ and $\pi_i\pi_f = +1$, where π_i and π_f are the parities

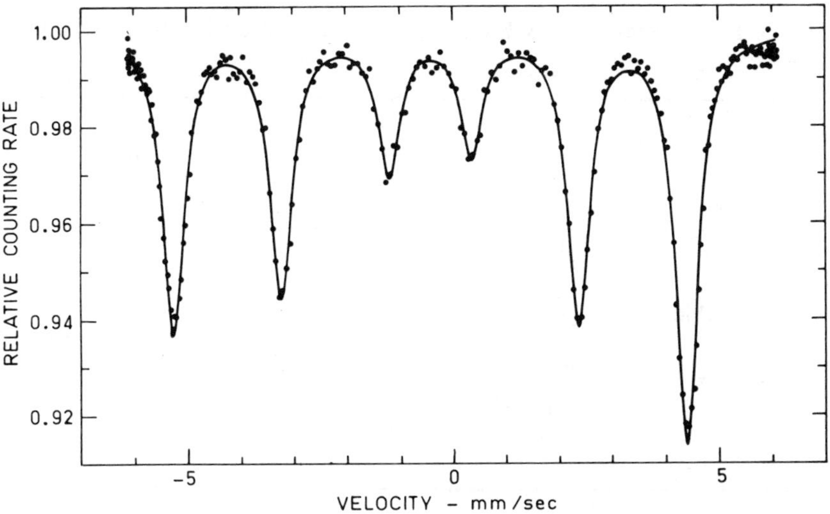

Fig. 13. Polarized spectrum of Co^{57} in palladium. The source was inside the mixing chamber, where the temperature of the surrounding liquid was 0.07°K. The stainless steel absorber was outside the cryostat at room temperature. The opposite sign of the hyperfine fields at Co^{57} in iron and palladium is clear when Figs. 12 and 13 are compared.

of the initial and final nuclear states, respectively. We will now discuss this problem in some more detail and also include first forbidden transitions with the selection rule $|I_i - I_f| = 0, 1$ and $\pi_i \pi_f = -1$.

It has been shown [20] that the Hamiltonian describing the β-decay interaction can be expanded into multipoles much the same way it is done in the case of γ-radiation. However, in β-decay it is generally not true that only the multipoles of lowest possible order have to be considered. Table II lists the nuclear matrix elements of different rank for allowed and first forbidden β-decay together with their selection rules. The population of a hyperfine level m_f in the daughter is given after integration over all directions and lepton spins by

$$P(m_f) = \sum_{m_i} P(m_i)\, W(m_i, m_f) \tag{12}$$

We separate the transition probability into a Clebsch–Gordan coefficient and a reduced matrix element [21]. For rank zero (scalar) we have

$$W^{(0)}(m_i, m_f) = C^{(0)}(m_i, m_f)\, |\langle \| S \| \rangle|^2 \tag{13}$$

Table II. Nuclear Matrix Elements for Allowed and First Forbidden β-Decay

Decay type	Matrix element	Tensor rank of the transition operator	Selection rules	
			$\lvert I_i - I_f \rvert$	$\pi_i\pi_f$
Allowed	$\int 1$	0	0	+1
	$\int \boldsymbol{\sigma}$	1	0,1; no $0 \leftrightarrow 0$	+1
First forbidden	$\int \gamma_5$	0	0	−1
	$\int (\boldsymbol{\sigma} \cdot \mathbf{r})$	0	0	−1
	$\int \mathbf{r}$	1	0,1; no $0 \leftrightarrow 0$	−1
	$\int \boldsymbol{\alpha}$	1	0,1; no $0 \leftrightarrow 0$	−1
	$\int (\boldsymbol{\sigma} \times \mathbf{r})$	1	0,1; no $0 \leftrightarrow 0$	−1
	$\int B_{ij}$	2	0,1,2; no $0 \leftrightarrow 0$, $\frac{1}{2} \leftrightarrow \frac{1}{2}$, $0 \leftrightarrow 1$	−1

where

$$C^{(0)}(m_i, m_f) = \delta_{m_i, m_f} \tag{14}$$

with δ being the Kronecker symbol. Equation (13) is valid for the matrix elements $\int 1$, $\int \gamma_5$ and $\int (\boldsymbol{\sigma} \cdot \mathbf{r})$. For rank one (vector) we have

$$W^{(1)}(m_i, m_f) = C^{(1)}(m_i, m_f)\, |\langle \| V \| \rangle|^2 \tag{15}$$

where

$$C^{(1)}(m_i, m_f) = |\langle j_f m_f 1 M | j_f 1 j_i m_i \rangle|^2 \tag{16}$$

is the square of the Clebsch–Gordan coefficient for dipolar radiation. Equation (16) is valid for the matrix elements $\int \boldsymbol{\sigma}$, $\int \mathbf{r}$, $\int \boldsymbol{\alpha}$, and $\int (\boldsymbol{\sigma} \times \mathbf{r})$. For rank two (tensor) we have

$$W^{(2)}(m_i, m_f) = C^{(2)}(m_i, m_f) |\langle \| T \| \rangle|^2 \tag{17}$$

where

$$C^{(2)}(m_i, m_f) = |\langle j_f m_f 2 M | j_f 2 j_i m_i \rangle|^2 \tag{18}$$

is the square of the Clebsch–Gordan coefficient for quadrupolar radiation. Equation (17) is valid for the matrix element $\int B_{ij}$.

In the allowed β-decay only the two matrix elements $\int 1$ and $\int \boldsymbol{\sigma}$ are present. They represent the Fermi and Gamow–Teller transitions, respectively. The pure Fermi transition ($I_i = I_f = 0$; $\pi_i\pi_f = +1$) and the mixed Fermi–Gamow–Teller transitions ($|I_i - I_f| = 0$; $\pi_i\pi_f = +1$) are of little interest here. A pure Gamow–Teller decay is found for several Mössbauer sources, in particular, for Co^{57}. In this case only the element $\int \boldsymbol{\sigma}$ is present and the ratio of two populations of the daughter nucleus, which we measure as the nuclear polarization effect [cf. Eqs. (8) and (9)], is simply given by the ratio of the appropriate coefficients $C^{(1)}(m_i, m_f)$. This calculation is explicitly performed in ref. [2] for the case of Co^{57}.

In a forbidden β-decay generally all six matrix elements will be present unless some of them are excluded by the selection rules. The relative strengths of $W^{(0)}$, $W^{(1)}$, and $W^{(2)}$ are not known *a priori*. However, the observation of the population of the different hyperfine levels in the daughter nucleus with the Mössbauer effect may give direct evidence as to the relative strengths of the different reduced matrix elements. It is customary to define mixing ratios of the reduced matrix elements in the following manner:

$$\delta^2 = \frac{|\langle \| V \| \rangle|^2}{|\langle \| S \| \rangle|^2} \quad \text{or} \quad \frac{|\langle \| T \| \rangle|^2}{|\langle \| V \| \rangle|^2} \tag{19}$$

By evaluating the Clebsch–Gordan coefficients we will be able to obtain the mixing ratios [18] from Mössbauer measurements on polarized sources which decay by first forbidden β-transitions. Two experiments of this type have been performed so far. By using a He^3 refrigerator the Munich group [22] has studied Tm^{169}, which is populated by the β-decay of Er^{169}. In the following we describe in more detail our measurements [23] on the $Pt^{197} \rightarrow Au^{197}$ decay.

The ground state of Pt^{197} (parent) and the 77-keV excited state of Au^{197} (daughter) both have a spin $I = \frac{1}{2}$ (cf. Fig. 11). In this case the polarization is defined as

$$\mathscr{P} = P(+\tfrac{1}{2}) - P(-\tfrac{1}{2}) \tag{20}$$

where P is the normalized level population as given by Eq. (4).

For obtaining a hyperfine splitting in our source we melted 1 at. % of enriched Pt^{196} into iron metal. The parent activity was created by neutron activation $Pt^{196}(n, \gamma)Pt^{197}$ of the alloy in the reactor of the Technical University of Helsinki. The iron matrix gave a negligible background to the 77-keV γ-line of Pt^{197}. A Mössbauer spectrum obtained with this source *vs.* a single-line gold metal absorber at 4.2°K is shown in Fig. 14. As expected the splitting parameters are equal to those found with an absorber of 1% gold

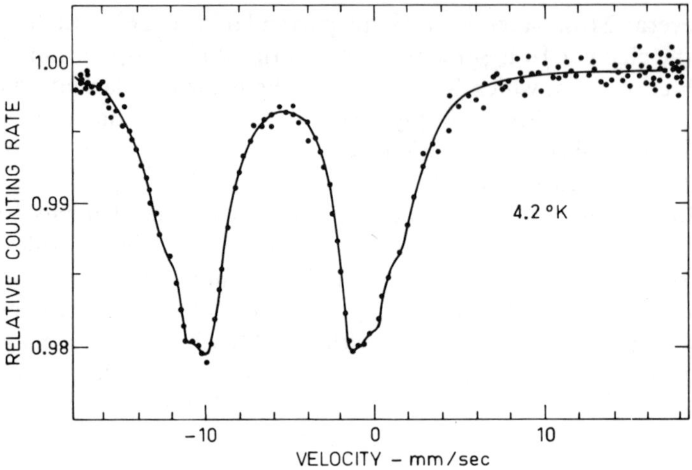

Fig. 14. Mössbauer spectrum of Au^{197} for a source of 1 at. % Pt^{197} in iron at 4.2°K. The absorber is metallic gold at 4.2°K.

alloyed into iron [24]. By cooling the source to 0.06°K we obtained the spectrum of Fig. 15. Nuclear polarization effects are evident as differences in the line intensities. From Fig. 11 it becomes clear that the right half of the spectrum (Fig. 15) contains lines which all originate from the $m = -1/2$ sublevel of the 77-keV state. Similarly, all lines on the left side originate from the $m = +1/2$ level. We denote with $S(-)$ and $S(+)$ the areas under the left and right sides of

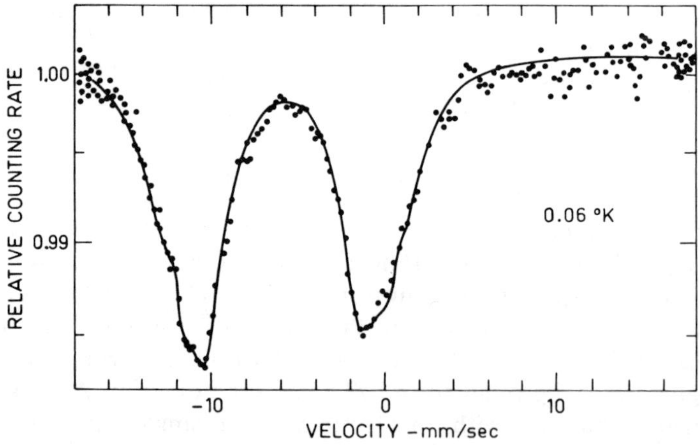

Fig. 15. Mössbauer spectrum of Au^{197} for a source of 1 at. % Pt^{197} in iron at 0.06°K. The absorber is metallic gold at 1.5°K. The polarization is clearly visible.

the spectrum, respectively. The polarization of the 77-keV state of Au^{197} can now be written [cf. Eq. (20)]:

$$\mathscr{P}_{Au} = \frac{S(+1/2) - S(-1/2)}{S(+1/2) + S(-1/2)} \tag{21}$$

The sublevel populations of the excited state of Au^{197} and the ground state of Pt^{197} are connected by a first forbidden β-transition with $\Delta I = 0$ and $\pi_i \pi_f = -1$. Since the transition occurs between two nuclear states $I_i = I_f = 1/2$, the tensor element will vanish according to its selection rule (Table II). We thus can write, using Eqs. (10) and (13–18).

$$\begin{aligned} P_{Au}(+1/2) = {} & P_{Pt}(+1/2)[C^{(0)}(+1/2, +1/2)|\langle\|S\|\rangle|^2 \\ & + C^{(1)}(+1/2, +1/2)|\langle\|V\|\rangle|^2] \\ & + P_{Pt}(-1/2)[C^{(0)}(-1/2, +1/2)|\langle\|S\|\rangle|^2 \\ & + C^{(1)}(-1/2, +1/2)|\langle\|V\|\rangle|^2] \end{aligned} \tag{22a}$$

and

$$\begin{aligned} P_{Au}(-1/2) = {} & P_{Pt}(+1/2)[C^{(0)}(+1/2, -1/2)|\langle\|S\|\rangle|^2 \\ & + C^{(1)}(+1/2, -1/2)|\langle\|V\|\rangle|^2] \\ & + P_{Pt}(-1/2)[C^{(0)}(-1/2, -1/2)|\langle\|S\|\rangle|^2 \\ & + C^{(1)}(-1/2, -1/2)|\langle\|V\|\rangle|^2] \end{aligned} \tag{22b}$$

The squares of the Clebsch–Gordan coefficients $C^{(L)}$ of Eq. (22) are listed in Table III. Using those and Eq. (20), we find:

$$\delta^2 = \frac{|\langle\|V\|\rangle|^2}{|\langle\|S\|\rangle|^2} = \frac{\mathscr{P}_{Pt} - \mathscr{P}_{Au}}{\frac{1}{3}\mathscr{P}_{Pt} + \mathscr{P}_{Au}} \tag{23}$$

The values of $\mathscr{P}_{Au}$ for different temperatures are obtained experimentally from spectra like Fig. 15, and are listed in the second column

Table III. Values of the Coefficients $C^{(L)}(m_i, m_f)$

	$L = 0$		$L = 1$	
m_i	$m_f = -\frac{1}{2}$	$m_f = +\frac{1}{2}$	$m_f = -\frac{1}{2}$	$m_f = +\frac{1}{2}$
$-\frac{1}{2}$	1	0	$\frac{1}{3}$	$\frac{2}{3}$
$+\frac{1}{2}$	0	1	$\frac{2}{3}$	$\frac{1}{3}$

of Table IV. The source polarization $\mathscr{P}_{Pt}$ can be calculated from the relation

$$\mathscr{P}_{Pt} = \tanh[\mu_{Pt}(H_i)_{Pt}/kT] \tag{24}$$

The hyperfine field of dilute platinum in iron is $(H_i)_{Pt} = -1.28$ MOe [25]. The ground state of Pt^{197} is formed by the odd neutron in a $p_{1/2}$ state. The same assignment is given to the ground state of Pt^{195}, for which a magnetic moment of $+0.606$ nm has been measured [26].

Table IV. Results of Polarization Measurements on Pt^{197}

T	$\mathscr{P}_{Pt}$, %	$\mathscr{P}_{Au}$, %	δ^2
20	±0	−0.2	—
4.2	±0	+0.2	—
0.120	−22.9	+2.8	5.3
0.091	−29.8	+3.0	4.7
0.088	−30.8	+3.6	5.2
0.080	−33.6	(+4.6)	(5.8)
0.060	−43.5	(+5.6)	(5.5)
	AVERAGE		5.3

Within the framework of the shell model the dipole moment is produced by the unpaired nucleon. In the absence of experimental measurements we estimate the moment of Pt^{197} as $\mu_{Pt} = (0.60 \pm 0.05)$ nm. With these values $\mathscr{P}_{Pt}$ has been calculated and is listed in the third column of Table IV. From $\mathscr{P}_{Au}$ and $\mathscr{P}_{Pt}$ we obtain δ^2, which is given in the last column of Table IV. Our final result is $\delta^2 = 5.3 \pm 0.8$. The error is rather large, due to the uncertainty in μ_{Pt}. However, the result shows clearly that the vector matrix elements play the dominant role in this β-decay.

In principle, $\mu_{Pt}(H_i)_{Pt}$ does not have to be known. If we rearrange Eq. (23) and use Eq. (24), we find

$$\mathscr{P}_{Au} = \tanh[\mu_{Pt}(H_i)_{Pt}/kT](1 - \tfrac{1}{3}\delta^2)/(1 + \delta^2) \tag{25}$$

In order to determine δ^2 and $\mu_{Pt}(H_i)_{Pt}$ simultaneously, measurements of $\mathscr{P}_{Au}$ should be made over a wide temperature range, particularly at very low temperatures, where the tanh saturates. The two parameters could then be calculated by fitting the experimental data to Eq. (25). So far it has not been possible to make measurements at low enough temperatures to employ this method. The Kapitza thermal boundary resistance, as discussed above, may effectively prevent these experiments.

INVESTIGATIONS OF MATERIALS WITH LOW TRANSITION TEMPERATURES

The widest use of a dilution refrigerator in conjunction with a Mössbauer spectrometer will probably be studies of materials with low transition temperatures. Of special interest are Curie or Néel points well below 1°K. A Japanese group made some detailed measurements [5] on the magnetic properties of ferrocyanide complexes, with the aid of an adiabatic demagnetization cryostat. We have investigated [16] divalent europium salts down to 0.04°K. A brief discussion of the results will be presented here as an example of this type of experiment.

Europium compounds occur in two different electronic configurations, as $Eu^{2+}(4f^7, {}^8S_{7/2})$ and as $Eu^{3+}(4f^6, {}^7F_0)$. The electronic shell in the ground state of trivalent europium thus carries no net angular momentum, and magnetic ordering is not expected. Divalent compounds have a half-filled $4f$ shell and a pure S ground state; magnetic ordering is therefore possible. Earlier studies of europeous compounds have been heavily concentrated on the chalcogenides (EuO, EuS, EuSe, and EuTe). Their transition temperatures are well above 4.2°K and they have a simple cubic (NaCl) structure.

We have investigated the magnetic ordering in the more-ionic europeous salts, for instance, in $EuSO_4$. The 21.6-keV resonance in Eu^{151} was employed. The source was $Sm_2{}^{151}O_3$, which emits a slightly broadened single line. The source temperature was about 2°K.

All compounds investigated showed a magnetic transition. The hyperfine field as a function of temperature followed roughly the $S = 7/2$ Brillouin curve up to about $0.85T_c$. By further increasing the temperature the resonance line narrowed only very gradually and remained broadened until about 2°K. This effect is thought to be caused by paramagnetic relaxation [27]. The magnetic transition temperatures of our samples were determined by extrapolating the $S_{7}/_{2}$Brillouin function to T_c. Our results, together with the measured

Table V. Hyperfine Parameters of Eu^{2+} Compounds from Mössbauer Data

Compound	Magnetic transition temperature, °K	Saturation magnetic hyperfine field, kOe	Quadrupole coupling (e^2qQ), MHz	Isomer shift,* mm/sec	Ref.
$EuSO_4$	0.43 ± 0.05	324 ± 2	+200 ± 20	−14.5 ± 0.1	This work
EuF_2	1.0 ± 0.1	333 ± 3	−30 ± 15	−14.3 ± 0.1	This work
$EuCl_2$	1.1 ± 0.2	315 ± 3	−145 ± 15	−14.1 ± 0.1	This work
$EuCO_3$	1.05 ± 0.1	319 ± 3	−58 ± 8	−14.0 ± 0.1	This work
EuC_2O_4	1.9 ± 0.1	337 ± 3	+70 ± 8	−13.9 ± 0.1	This work
$EuC_2O_4 \cdot H_2O$	2.8 ± 0.1	336 ± 3	+60 ± 8	−13.9 ± 0.1	This work
EuTe	9	242	0	−12.9	[28, 29]
EuSe	4.8 ± 0.1	286 ± 3	0	−12.7	[29, 30]
EuS	16.5 ± 0.1	332 ± 3	< 8	−12.5 ± 0.1	This work
EuO	69	300 ± 8	0	−11.9	[29, 31]
Eu metal	89.6	265 ± 10	0	−8.3 ± 1	[31]
(Eu) CaF_2	—	341†	0	−13.2	[32]

* *Versus* Eu^{3+} in Sm_2O_4.
†From EPR data [33].

hyperfine parameters, are listed in Table V; several earlier measurements by other authors on the chalcogenides are also included.

Some of our compounds (e.g., $EuCl_2$) showed a contamination with Eu^{3+}; the europic resonance line remained unaffected even at the lowest experimental temperature of 0.04°K. This indicates that the electronic ground state of Eu^{3+} is well described by $J = 0$ and that admixtures from the $J = 1$ first excited state are very small. The data of Table V show no apparent correlation between hyperfine field and the isomer shift as has been proposed by Wickman *et al.* [31]. Most of the europeous compounds have saturation fields of about 325 kOe; this is about 5% less than the paramagnetic hyperfine field of 341 kOe found for Eu^{2+} in CaF_2 by EPR techniques [32]. The largest negative isomer shift was measured for $EuSO_4$. This compound thus presumably has the most-ionic bond [29]; however, it does not show the highest field value.

The crystal symmetry of the ionic europeous compounds is low, and therefore quadrupolar interaction should be present. For a transition element the electric field gradient at the nucleus in an ionic compound is given by

$$q \sim q_L(1 - \gamma_\infty) + q_i(1 - R_q) \tag{26}$$

The first term represents the field gradient q_L produced by the electric charges in the lattice, amplified by the Sternheimer antishielding factor $(1 - \gamma_\infty)$. The second term, the gradient caused by the alignment of an unfilled inner shell, vanishes for the electronic ground states of both Eu^{2+} and Eu^{3+}.

In spite of the fact that $(1 - \gamma_\infty) \approx 80$, the resulting field gradient at the europium nuclei is rather small and gives rise only to a slight broadening of the resonance line in the paramagnetic region [34, 51]. In the presence of a magnetic hyperfine splitting a small quadrupole interaction can be more easily observed as an asymmetry of the resonance lines. Figure 16 illustrates the Mössbauer spectra of $EuSO_4$ and of $EuCl_2$ at 0.054°K. By comparing these with Fig. 9, showing the spectrum of cubic EuS under similar experimental conditions, the presence of a quadrupolar shift in the ionic compounds of low symmetry is quite apparent.

Our data can thus be used for obtaining the ratio of the quadrupole moments of the excited and ground states of Eu^{151}, Q_e/Q_g. For this purpose the spectra have been analyzed with a Hamiltonian for an axially symmetrical quadrupole interaction ($\eta = 0$), and with the principal axis of the electric field gradient along H_i ($\theta = 0$). In view of

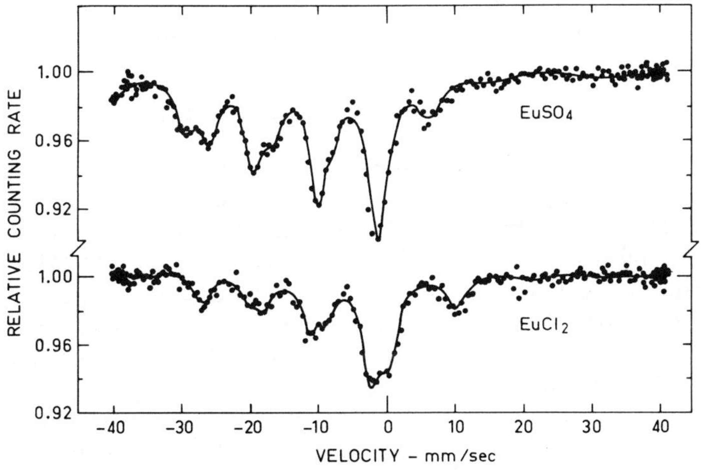

Fig. 16. Mössbauer spectra of Eu^{151} for $EuSO_4$ and $EuCl_2$ absorbers at 0.054°K. Several magnetic hyperfine lines are split by additional quadrupole interaction. The line at zero velocity in $EuCl_2$ is caused by a trivalent impurity. The difference in the two spectra is caused by the opposite sign of e^2qQ in the two compounds.

the rather low crystal symmetry in the compounds investigated, these assumptions may be an oversimplification. For studying the influence of the off-diagonal terms we have simulated spectra for different values of η and θ. Since $gH_i \gg 3e^2qQ/4I(2I - 1)$, the value of η does not influence the line positions within our resolution, and hence can be neglected. The introduction of an angle θ between q and H_i mainly changes the interaction energy e^2qQ, but has little influence on the ratio of the quadrupole moments Q_e/Q_g. Our results are summarized in Table VI, leading to $Q_e/Q_g = 1.30 \pm 0.05$. This value is in excellent agreement with a recent measurement on europium iron garnet [35], but differs from other current values as listed in Table VI. With

Table VI. Ratio of the Quadrupole Moments Q_e/Q_g of the Ground State and the 21.6-keV Excited State of Eu^{151}

Compound	Q_e/Q_g	Method
$EuSO_4$	1.28 ± 0.05	—
EuC_2O_4	1.23 ± 0.08	—
$EuCl_2$	1.35 ± 0.06	—
$EuCO_3$	1.32 ± 0.09	—
Weighted average	1.30 ± 0.05	Mössbauer
Stachel *et al.* [35]	1.28 ± 0.05	Mössbauer
Nozik *et al.* [34]	1.64 ± 0.07	Mössbauer
Carrigan *et al.* [36]	1.56 ± 0.25	Muonic X-rays

$Q_g = 1.16$ b [26] we find $Q_e = 1.51$ b. The quadrupole interactions in several of our europeous compounds, particularly in $EuSO_4$ (cf. Table V), are substantially larger than observed for the two sites in europium iron garnet [35]: (15 ± 6) MHz and (23 ± 6) MHz.

The 21.6-keV transition in Eu^{151} can be described as a proton-hole excitation ($2d_{5/2} \rightarrow 1g_{7/2}$). The corresponding proton–particle excitation is found in the 37.2-keV transition of Sb^{121}, which is also a well-established Mössbauer nucleus. The astonishing similarities of Eu^{151} and Sb^{121} are evident from Table VII. The quadrupole moments Q_g and Q_e of Sb^{121} can well be accounted for by the shell model [39], whereas for Eu^{151} one finds experimentally for Q_g roughly three times the calculated value. This is a clear indication of the onset of strong nuclear deformation near $A = 151$.

Table VII. Comparison of Nuclear Parameters of Sb^{121} and Eu^{151}

Parameter	Isotope	
	$_{51}Sb^{121}$	$_{63}Eu^{151}$
Ground state	$2d_{5/2}$	$2d_{5/2}$
Excited state	$1g_{7/2}$	$1g_{7/2}$
Transition	Proton–particle, 37.2 keV	Proton–hole, 21.6 keV
μ_g	+3.36 nm [26]	+3.46 nm [26]
g_e/g_g	+0.525 [37]	+0.533 [38]
Q_g (calc.)	−0.26 b [39]	+0.36 b [39]
Q_g (exp.)	−0.26 b [26]	+1.16 b [26]
Q_e/Q_g (exp.)	1.36 [37]	1.30 [this work]
$\Delta\langle r^2\rangle/\langle r^2\rangle$ (calc.)	$+0.16 \times 10^{-4}$ [41]	-1.5×10^{-4} [42]
$\Delta\langle r^2\rangle/\langle r^2\rangle$ (exp.)	-16×10^{-4} [37]	$+11 \times 10^{-4}$ [40]

The change in the nuclear charge radius $\Delta\langle r^2\rangle/\langle r^2\rangle$ with the 21.6-keV excitation has been discussed in detail by Hüfner *et al.* [43] and by Steichele [42]. According to their analysis we may write

$$\Delta\langle r^2\rangle_{\text{obs}} = \Delta\langle r^2\rangle_{\text{coll}} + \Delta\langle r^2\rangle_{\text{part}} \tag{27}$$

where "coll" and "part" refer to the collective and single-particle contributions, respectively. The collective term could be caused by a change in deformation, and may be expressed as

$$\Delta\langle r^2\rangle_{\text{coll}} = \Delta\langle r^2\rangle_{\text{def}} = (Q_{0,e}{}^2 - Q_{0,g}{}^2)/4Z^2\langle r^2\rangle \tag{28}$$

where $Q_{0,e}$ and $Q_{0,g}$ are the intrinsic quadrupole moments. We may consider the ground state and the first excited state of Eu^{151} as the lowest states of a 5/2 or a 7/2 rotational band. We then obtain, by using the projection formula for strong coupling [44] and the values $Q_g = 1.16$ b and $Q_e = 1.51$ b,

$$Q_{0,g} - Q_{0,e} = (3.25 - 3.23)\text{ b} \approx 0 \tag{29}$$

Thus, only the single-particle contribution to $\Delta\langle r^2\rangle$ should remain. A calculation by Steichele [42] gives $(\Delta\langle r^2\rangle/\langle r^2\rangle)_{\text{part}} = -1.50 \times 10^{-4}$, a value which is too small and of opposite sign. The same dilemma exists for Sb^{121}, where Uher *et al.* [41] obtain for $\Delta\langle r^2\rangle$ of a single-proton transition also the wrong sign and magnitude (cf. Table VII).*

*A very recent attempt to calculate the charge-radius change with the help of Migdal theory gave $\Delta\langle r^2\rangle/\langle r^2\rangle = +4.5 \times 10^{-4}$ for Eu^{151}, in satisfactory agreement with the experiment [51].

SOME FURTHER EXPERIMENTS AT VERY LOW TEMPERATURES

In this last section we shall briefly discuss some experiments for which the very low temperatures obtainable with a dilution refrigerator are of considerable help. Several of these measurements are planned for or are in progress at the Technical University of Helsinki.

Pseudoquadrupolar Interaction

This effect has been observed [45] in $Tm_2(SO_4)_3 \cdot 8H_2O$ and in $TmCl_3 \cdot 6H_2O$. In both of these compounds the electronic ground state and the first excited state are singlets and closely spaced together. By expanding the magnetic hyperfine interaction with the use of second-order perturbation theory we find

$$E_H = \langle g|A\mathbf{I}\cdot\mathbf{J}|g\rangle - |\langle g|A\mathbf{I}\cdot\mathbf{J}|e\rangle|^2/(E_e - E_g) + \ldots \qquad (30)$$

where g and e refer to the electronic ground and excited states, respectively, and A is the hyperfine interaction constant. Since both states are singlets, the first term in Eq. (30) vanishes. The second term, however, is large in these salts, owing to the small separation of the two states. It is called a "pseudoquadrupolar term," since it is proportional to m_I^2. The pseudoquadrupolar interaction can be observed in the spectra of the thulium salts mentioned above as shift of the quadrupole hyperfine lines at temperatures below 4°K. The center of gravity of the spectrum is not a constant, but shows a temperature dependence,

$$G(T) \propto \tanh[(E_e - E_g)/2kT] \qquad (31)$$

It is obvious that when $T \to 0$ one can measure the electronic level separation $E_e - E_g$, which so far has not been determined. This requires temperatures below 0.3°K.

Relaxation Measurements

In recent years the Mössbauer effect has become a very valuable tool for investigating relaxation phenomena, particularly paramagnetic relaxation. Various aspects of these measurements have been discussed in detail by Wickman [27].

The electronic spin fluctuations are caused by the spin-lattice and the spin-spin interactions. The spin-lattice relaxation, which dominates in magnetically dilute specimens, is determined by the coupling between the spins and the phonons. When a spin flips the excitation energy has to be transferred to the phonon bath. The probability of a spin flip associated with the excitation of a single phonon is proportional to the temperature (Kronig–Van Vleck, or direct process). Multiphonon excitations (Raman and Orbach processes) have a much stronger temperature dependence. For Kramer's doublets in the absence of a magnetic field we may write [46]

$$1/\tau = AT + BT^7 + C\exp[-(E_e - E_g)] \tag{32}$$

where τ is the relaxation time and A, B, and C, are constants. The different terms in this equation represent direct, Raman, and Orbach processes, respectively.

With decreasing temperature the relaxation time will become longer, and it can thus be observed with the Mössbauer effect. The direct Van Vleck process will dominate at low temperatures. The detailed temperature dependence of τ is complex, but when investigated gives valuable data both on the phonon spectrum and on the phonon-spin coupling. More information can be obtained from a simultaneous measurement of the temperature and of the magnetic field dependence of the relaxation time.

First-order theory predicts that the spin–spin relaxation time is independent of temperature, but the inclusion of higher-order interactions alters this simple picture. Investigations of spin–spin relaxation at very low temperatures should therefore be of considerable interest.

In rare-earth ions the electronic magnetic moment is produced by the $4f$ electrons. In many cases this inner electron shell is only weakly coupled to the lattice. Rather long relaxation times are thus observed even at 10–20°K. The actinide elements form the $5f$ transition series. There is considerable experimental evidence that the $5f$ electrons are much less shielded compared to the $4f$ electrons in the rare earths. Lower temperatures will thus be needed for investigations of relaxation phenomena in the actinides, which possess a number of rather favorable Mössbauer transitions. Radioactive heating will, however, be a serious experimental problem.

Of particular interest should be relaxation phenomena in conducting media at very low temperatures. Mössbauer effect measurements will provide a valuable tool for the investigation of processes like the Koringa relaxation in metals, where data using other resonance techniques are only difficult to obtain.

Localized Magnetic Moments

If small amounts of transition elements are substituted as impurities into nonmagnetic metallic hosts, localized magnetic moments are usually created. When the impurity concentration is low enough the material will be paramagnetic, and we will have a group of mutually independent magnetic moments. By applying an external field H, the moments can be partially oriented to give an average component

$$\langle \mu \rangle = g\mu_\beta \langle J_z \rangle = \mu B_J(g\mu_\beta JH/kT) \tag{33}$$

in the direction of H. Here g is the spectroscopic splitting factor, μ_β is the Bohr magneton, J is the electronic angular momentum, and $B_J(x)$ is the Brillouin function

$$B_J(x) = \frac{2J+1}{2J}\coth\left[\frac{2J+1}{2J}x\right] - \frac{1}{2J}\coth\left[\frac{x}{2J}\right] \tag{34}$$

with $x = g\mu_\beta JH/kT$. The average moment given in Eq. (33) will, in turn, produce an internal field

$$H_i(T) = H_i(0)B_J(x) \tag{35}$$

The change of $H_i(T)$ with temperature for small values of H/T becomes

$$H_i(T)/H_i(0) \approx (J+1)\mu H/3JkT \tag{36}$$

which provides a measure of $(J+1)\mu/3J$.

The detailed approach of $H_i(T)$ to saturation depends on the angular momentum J. Accurate experimental data on H_i over a wide temperature range will thus permit a determination of μ and J. Their combination determines g. In favorable cases this method may prove to be a good way for measuring electronic moments. Since the degree of orientation depends on H/T, very low temperatures are of considerable help for approaching saturation.

As an example, we mention the iron-in-palladium system, which has been studied by Kitchens *et al.* [47, 48]. They found a very high magnetic moment, $\mu = 12.6\mu_\beta$, in agreement with previous magnetic measurements [49], and an angular momentum $J = 6$. Palladium has 0.6 electron deficiency in its 5*d* band and is a highly polarizable host. It is assumed that approximately $3\mu_\beta$ of the observed total magnetic moment is at each iron atom, with the surrounding polarized palladium matrix extending to about 10 Å from the impurity. The total amount of magnetic moment associated with palladium atoms within this sphere is thus 9–$10\mu_\beta$.

Localized magnetic moments and the associated bound states (Kondo effect) have been under intensive theoretical and experimental study during the last few years. Application of the Mössbauer effect for these investigations makes it possible to use very low concentrations (10^{-1}–10^{-4} at. %) of impurity atoms, thus preventing a mutual interaction of the localized moments. Systems most readily accessible to theoretical calculations are those in which a transition-metal impurity has been introduced into a noble metal.

The formation of a bound state between the localized magnetic impurity spins and the conduction electron spins will cause a change in the saturation hyperfine field, due to the change in total electronic moment. This effect has been investigated with the Mössbauer technique by Frankel *et al.* [50] on a sample of 10^{-4}–10^{-3} at. % of Co^{57} diffused into copper. Their measurements show that the bound state can be partially broken up by a large external magnetic field.

There are several cases in which studies of localized moments and bound states have shown unexpected effects. It would certainly be advantageous to extend these measurements to considerably lower temperatures.

ACKNOWLEDGMENTS

The work described in this paper has been performed in collaboration with G. J. Ehnholm, P. Reivari, and G. K. Shenoy. Financial support from the Jenny and Antti Wihuri Foundation is gratefully acknowledged. One of us (G.M.K.) would like to thank the Department of Technical Physics of the Technical University of Helsinki for their hospitality. We are also much indebted to A. Freeman for many suggestions. Discussions with P. Kienle and B. D. Dunlap were very helpful to us. H. Pink from Siemens & Halske A.G., Munich, supplied the divalent europium salts used in this investigation.

REFERENCES

1. J. G. Dash, R. D. Taylor, P. P. Craig, D. E. Nagle, D. R. Cochran, and W. E. Keller, *Phys. Rev. Letters* **5**:152 (1960).
2. J. G. Dash, R. D. Taylor, D. E. Nagle, P. O. Craig, and W. M. Visscher, *Phys. Rev.* **122**:1116 (1961).
3. R. D. Taylor, in: *Temperature, Its Measurement and Control in Science and Industry*, C. M. Herzfeld and F. G. Brickwedde, eds. (Reinhold Publishing Corp., New York, 1962), Vol. 3, Part 1, p. 139.
4. R. D. Taylor, in: *The Mössbauer Effect* (Proc. Second International Conference on the Mössbauer Effect), D. M. Compton and A. H. Schoen, eds. (John Wiley and Sons, New York, 1962), p. 203.
5. M. Shinohara, A. Ito, T. Fujita, A. Ishigaki, and K. Ôno, *Japan J. Appl. Phys.* **6**:982 (1967); **7**:170 (1968).
6. H. E. Hall, P. J. Ford, and K. Thompson, *Cryogenics* **6**:80 (1966).
7. B. S. Neganov, N. Borisov, and M. Liburg, *Soviet Phys.–JETP* **23**:959 (1966).
8. O. E. Vilčhes and J. C. Wheatley, *Physics Letters* **24A**:440 (1967); **25A**:344 (1967).

9. J. C. Wheatley, O. E. Vilches, and W. R. Abel, *Physics* **4**:1 (1968).
10. O. V. Lounasmaa, *Helv. Phys. Acta* **41**:1021 (1968).
11. G. M. Kalvius, in: *Mössbauer Effect Methodology*, Vol. 1, I. J. Gruverman, ed. (Plenum Press, New York, 1965), p. 163ff.
12. W. A. Steyert and M. D. Daybell, in: *Mössbauer Effect Methodology*, Vol. 4, I. J. Gruverman, ed. (Plenum Press, New York, 1965), p. 3ff.
13. G. J. Ehnholm, T. E. Katila, O. V. Lounasmaa, and P. Reivari, *Cryogenics* **8**:136 (1968).
14. L. D. Roberts and J. O. Thomson, *Phys. Rev.* **129**:664 (1963).
15. G. J. Ehnholm, T. E. Katila, O. V. Lounasmaa, and P. Reivari, *Physics Letters* **25**A:758 (1967).
16. J. G. Ehnholm, T. E. Katila, O. V. Lounasmaa, P. Reivari, and G. M. Kalvius, *Proc. Eleventh International Conference on Low Temperature Physics*, Vol. 1, J. F. Allen, D. M. Finlayson, and D. M. McCall, eds. (St. Andrews, Scotland, 1968), p. 528.
17. S. Margulies and J. R. Ehrman, *Nucl. Instr. Meth.* **12**:131 (1961).
18. G. A. Bykov and Pham Zuy Hien, *Soviet Phys.—JETP* **16**:646 (1963).
19. P. Reivari, *Phys. Rev. Letters* **22**:167 (1969).
20. H. A. Weidenmuller, *Rev. Mod. Phys.* **33**:574 (1961).
21. S. R. DeGroot, H. A. Tolhoek, and W. Huiskamp, in: *Alpha- Beta- and Gamma-Ray Spectroscopy*, K. Siegbahn, ed. (North-Holland Publishing Co., Amsterdam, 1968), p. 1199ff.
22. R. Geller, F. Wagner, W. Wiedemann, and P. Kienle, private communication (1968).
23. G. M. Kalvius, G. J. Ehnholm, T. E. Katila, O. V. Lounasmaa, and P. Reivari, *Bull. Am. Phys. Soc.* **13**:1469 (1968).
24. R. L. Cohen, *Phys. Rev.* **171**:343 (1968).
25. D. A. Shirley, in: *Hyperfine Structure and Nuclear Radiations*, E. Mathias and D. A. Shirley, eds. (North-Holland Publishing Co., Amsterdam, 1968), p. 979ff.
26. V. S. Shirley, in: *Hyperfine Structure and Nuclear Radiations*, E. Mathias and D. A. Shirley, eds. (North-Holland Publishing Co., Amsterdam, 1968), p. 985.
27. H. H. Wickman, in: *Chemical Applications of Mössbauer Spectroscopy*, V. I. Goldanskii and R. H. Herber, eds. (Academic Press, New York, 1968), p. 548ff.
28. R. L. Cohen, *Bull. Am. Phys. Soc.* **13**:667 (1968).
29. G. Gerth, P. Kienle, and K. Luchner, *Physics Letters* **27**A:557 (1968).
30. F. Holtzberg, G. M. Kalvius, G. K. Shenoy, and B. D. Dunlap, unpublished (1968).
31. H. H. Wickman, I. Nowik, J. H. Wernick, D. A. Shirley, and R. B. Frankel, *J. Appl. Phys.* **37**:1246 (1966).
32. H. Maletta, W. Heidrich, and R. L. Mössbauer, *Physics Letters* **25**A:295 (1968).
33. J. M. Baker, F. I. Williams, *Proc. Roy. Soc.* (*London*) **A267**:283 (1963).
34. A. J. Nozik, M. Kaplan, and A. I. Weiss, *Bull. Am. Phys. Soc.* **13**:894 (1968).
35. M. Stachel, S. Hüfner, G. Crecelius, and D. Quitmann, *Physics Letters* **28**A:188 (1968).
36. R. A. Carrigan, Jr., P. D. Gupta, R. B. Sutton, M. Suzuki, A. C. Thompson, R. E. Coté, W. V. Prestwich, A. K. Gaigalas, and S. Raboy, *Phys. Rev. Letters* **20**:874 (1968).
37. S. L. Ruby, in: *Mössbauer Effect Methodology*, Vol. 3, I. J. Gruverman, ed., (Plenum Press, New York, 1967), p. 203ff.
38. P. Kienle, *Rev. Mod. Phys.* **36**:372 (1964).
39. H. Horie and A. Arima, *Phys. Rev.* **99**:778 (1955).
40. I. Koniordos and R. Winkler, *Phys. Letters* **27**A:198 (1968).
41. R. A. Uher, private communication; R. A. Uher and R. A. Sorensen, *Nucl. Phys.* **86**:1 (1966).
42. E. Steichele, *Z. Physik* **201**:331 (1967).
43. S. Hüfner, P. Kienle, D. Quitmann, and P. Brix, *Z. Physik* **187**:67 (1965).
44. A. Bohr and B. R. Mottelson, *Mat. Fys. Medd. Dan. Vid. Selsk.* **27**(16) (1953).

45. M. J. Clauser, E. Kankeleit, and R. L. Mössbauer, *Phys. Rev. Letters* **17**:5 (1966).
46. R. Orbach, in: *Fluctuation, Relaxation and Resonance in Magnetic Systems*, D. Ter Haar, ed. (Oliver and Boyal, London, 1962), p. 219ff.
47. T. A. Kitchens and P. P. Craig, *J. Appl. Phys.* **37**:1187 (1966).
48. T. A. Kitchens, W. A. Steyert, and R. D. Taylor, *Phys. Rev.* **138**:A467 (1965).
49. A. M. Clogston, B. T. Mathias, M. Peter, H. J. Williams, E. Corenzwit, and R. C. Sherwood, *Phys. Rev.* **125**:541 (1962).
50. R. B. Frankel, N. A. Blum, B. B. Schwartz, and D. J. Kim, *Phys. Rev. Letters* **18**:1051 (1967).
51. J. Speth, Thesis, Technical University, Munich, 1968; *Phys. Letters* **28B**:625 (1969).

Index